"This fine monograph successfully captures the complexities of the geopolitics of Irish–British relations and nation- and state-making in Ireland. It explores the diverse ways in which political territoriality finds expression in the mutation of ethnonational territorial conflict, the making of internal and external borders, and the crafting of ethnic, national, and sectarian identities as place identities. Adopting a cross-scalar geographic perspective on Ireland's geopolitics that spans the region, cities, and their granular public spaces, Dempsey pays close attention to how place and landscape manifest fundamental elements of conflict, nationalism, and peacebuilding. Based on significant fieldwork and archival research, the book makes important contributions to political and cultural geography. Ultimately, Dempsey's erudite narrative helps us understand the intricate geopolitical transformation of Ireland at key stages of state formation, after the region's Europeanist turn and, now, in post-Brexit's uncertain times. All that is done in a manner accessible to a general audience while providing sophisticated scholarly analysis that would satisfy a broad range of academic readers."

Alex G. Papadopoulos, Department of Geography, DePaul University, USA

"This book is essential reading for anyone interested in understanding the complexity of Ireland's political geography over the past 100 years. As a decade of centenary commemorations draw to a close, Dempsey's book is particularly timely. Reflecting on seminal moments from partition to Brexit, 'An Introduction to the Geopolitics of Conflict, Nationalism, and Reconciliation in Ireland' explores the role of geography in ethnonational conflict and identity politics. Beautifully written and expertly framed, it offers a compelling insight into the trajectory of conflict and peacebuilding on this island."

Sara McDowell, School of Geography and Environmental Sciences, Ulster University

"This book is an informative, accessible, engaging introduction to the cultural and political geography of Ireland. Drawing on extensive fieldwork and the voices of local people, Kara E. Dempsey has put together an account that offers telling insights into the ways geopolitical context reflects and shapes ethnonational dynamics in the country. The book would be an excellent resource for courses in Irish Studies, and a valuable complement to more general readings in political geography and European Studies classes."

Alec Murphy, Department of Geography, University of Oregon, USA

AN INTRODUCTION TO THE GEOPOLITICS OF CONFLICT, NATIONALISM, AND RECONCILIATION IN IRELAND

This book examines ethnoterritorial conflict and reconciliation in Ireland from the 1916 Rising to Brexit (2021), including the production and consequences of the island's two distinct political units.

Highlighting key geographic themes of bordering, unity, division, and national narratives, it explores how geopolitical space has been employed over time to (re)define divided national allegiances throughout Ireland and within Irish–British relations. The analysis draws from in-depth interviews and archival research, and spans supranational, state, municipal, neighborhood, and individual scales. The book pays particular attention to uneven power structures, statecraft, perceived truths, lived experiences, reconciliation efforts, and renegotiations of national narratives in the production of symbolic landscapes, divided cities, and "shared" space. *An Introduction to the Geopolitics of Conflict, Nationalism, and Reconciliation in Ireland* provides readers with an analysis of geopolitical power relations and different spatial productions of conflict and peacebuilding in Ireland.

Offering deeper understanding of these historic and contemporary geopolitical intersections, this book makes a valuable contribution to the fields of Political Geography, Border Studies, Irish Studies, European Studies, International Relations, Cultural Geography, and Regional Studies.

Kara E. Dempsey is Associate Professor of political geography at Appalachian State University, part of the University of North Carolina system, and former Director of Irish Studies at DePaul University. She studies the intersection of politics and geography at various scales, particularly the utilization of space in the negotiation and production of ethnonational conflicts, consolidation of state and regional power, conflict transformation, and peacebuilding processes.

Routledge Geopolitics Series

Series Editors

Klaus Dodds

Professor of Geopolitics at the Department of Geography,
Royal Holloway University of London, Egham Surrey UK.
k.dodds@rhul.ac.uk

Reece Jones

Professor of Geography at the Department of Geography,
University of Hawai'i at Manoa, Hawai'i, USA.
reecej@hawaii.edu

Geopolitics is a thriving area of intellectual enquiry. The *Routledge Geopolitics Series* invites scholars to publish their original and innovative research in geopolitics and related fields. We invite proposals that are theoretically informed and empirically rich without prescribing research designs, methods, and/or theories. Geopolitics is a diverse field making its presence felt throughout the arts and humanities, social sciences, and physical and environmental sciences. Formal, practical, and popular geopolitical studies are welcome as are research in areas informed by borders and bordering, elemental geo-politics, feminism, identity, law, race, resources, territory and terrain, materiality, and objects. The series is also global in geographical scope and interested in proposals that focus on past, present, and future geopolitical imaginations, practices, and representations.

As the series is aimed at upper-level undergraduates, graduate students, and faculty, we welcome edited book proposals as well as monographs and textbooks which speak to geopolitics and its relationship to wider human geography, politics and international relations, anthropology, sociology, and the interdisciplinary fields of social sciences, arts, and humanities.

Published

Hellenic Statecraft and the Geopolitics of Difference
Edited by Alex G. Papadopoulos and Triantafyllos G. Petridis

An Introduction to the Geopolitics of Conflict, Nationalism, and Reconciliation in Ireland
Kara E. Dempsey

For more information about this series, please visit: www.routledge.com/Routledge-Geopolitics-Series/book-series/RFGS

AN INTRODUCTION TO THE GEOPOLITICS OF CONFLICT, NATIONALISM, AND RECONCILIATION IN IRELAND

Kara E. Dempsey

LONDON AND NEW YORK

Cover image: © Getty Images

First published 2023
by Routledge
4 Park Square, Milton Park, Abingdon, Oxon OX14 4RN

and by Routledge
605 Third Avenue, New York, NY 10158

Routledge is an imprint of the Taylor & Francis Group, an informa business

British Library Cataloguing-in-Publication Data
A catalogue record for this book is available from the British Library

Library of Congress Cataloging-in-Publication Data
Names: Dempsey, Kara E., author.
Title: An introduction to the geopolitics of conflict, nationalism, and reconciliation in Ireland / Kara E. Dempsey.
Description: Abingdon, Oxon ; New York, NY : Routledge, 2022. | Includes bibliographical references and index.
Identifiers: LCCN 2022002523 (print) | LCCN 2022002524 (ebook)
Subjects: LCSH: Ireland—Politics and government—1901–1910. | Ireland—Politics and government—1922– | Ireland—History—1901–1910. | Ireland—History—1922– | Geopolitics—Ireland—History. | Nationalism—Ireland—History. | Conflict management—Ireland—History. | Home rule—Ireland. | Ireland—History—Autonomy and independence movements. | Irish question.
Classification: LCC DA963 .D496 2022 (print) | LCC DA963 (ebook) | DDC 320.9415/0904—dc23/eng/20220504
LC record available at https://lccn.loc.gov/2022002523
LC ebook record available at https://lccn.loc.gov/2022002524

ISBN: 978-0-367-69265-0 (hbk)
ISBN: 978-0-367-69266-7 (pbk)
ISBN: 978-1-003-14116-7 (ebk)

DOI: 10.4324/9781003141167

Typeset in Bembo
by Apex CoVantage, LLC

This book is dedicated to my family, who share my love of Ireland.

CONTENTS

FIGURES

MAPS

TABLES

PREFACE

The initial idea for this book took shape while I was teaching undergraduate and graduate-level Human Geography and Irish Studies courses that focused on the partition of Ireland and territorialized national identities on the island. Students in these courses were often curious about the origins and impacts of these political divisions, as well as historical and contemporary reconciliation efforts that mark the stories of the Republic of Ireland and the United Kingdom.

To better contextualize the formation of various political entities, contested claims to territory, and the construction of national identities in Ireland, this book focuses on events from 1916 to 2021. It explores interrelated sites and scales of geographies of citizenship, constructions and manipulations of place-based nationalisms, the layered and contested meaning(s) of space, the (re)production of segregated geographies, and spatialized reconciliation processes on the island. The nature of this investigation lends itself to imbricating discussions of specific time periods and key themes to provide a more comprehensive examination of the role of geography in relation to conflict and reconciliation in Ireland. While each state expresses its unique nexus of geography, culture, economy, and political reality, there are lessons from Ireland's history with global relevance, and stories that underscore commonalities of the practices of statecraft, bordering, and formation of national identities. In this way, the lessons learned in the study of Ireland are applicable beyond the borders in Ireland.

Today, Ireland is at another critical juncture. For example, Brexit is particularly relevant throughout Ireland as a result of border (re)negotiations and because of the European Union's financial underpinning of many initiatives for reconciliation and fostering community cohesion in Northern Ireland. Ireland faces other significant transformative geopolitical and cultural junctures regarding the peace process in Northern Ireland (which remains dynamic and subject to both progress and regression), international migration in Europe, and geopolitical relationships among the Republic of Ireland, the United Kingdom, and the European Union. I hope that this book's critical investigations of the aforementioned will provide insight into the role of geography in conflict, nationalism, and reconciliation efforts in Ireland and worldwide.

ACKNOWLEDGMENTS

This book would not have been possible without the help and support of many of my friends and colleagues, including Bob Ostergren, Colin Flint, Win Curran, Natalie Koch, Sara McDowell, Orhon Myadar, Lindsay Naylor, the wonderful Allie Shay and Elizabeth Shay for going *far* above and beyond, Faye Leerink, (Editor), Ramachandran Vijayaragavan (Project manager), Saskia van de Gevel, Harrison Brown, Emily Dempsey, Mary Dempsey, Bobbi Schrank, Jane Anderson, Becky Cherkasky, Octavia Barefoot, Maggie Sugg, Liz McGoey, Jess Boll, Lauren Andersen, Alex Papadopoulos, Euan Hague, the truly inspirational Rev. Bill Shaw OBE, 174 Trust, and the kind people of Ireland.

I also would like to especially thank my family – mom, dad, Conor, Liz, Steph, Piper, F.G. and U.R., Jonathan, and Maisy – for their endless support of this project.

ABBREVIATIONS

AIA	Anglo-Irish Agreement, 1985
B Specials	Ulster Special Constabulary, auxiliaries under the RUC
C na mB	(Cumann na mBan) Women's paramilitary organization
Dáil	(Dáil Eireann) Irish Parliament
DUP	Democratic Unionist Party
EEC	European Economic Community, precursor to EU
EU	European Union
FF	Fianna Fail
FG	Fine Gael
GAA	Gaelic Athletic Association
ICA	Irish Citizen Army
INLA	Irish National Liberation Army
IPP	Irish Parliamentary Party
IRA	Irish Republican Army
IRB	Irish Republican Brotherhood
MP	Member of Parliament
NFE	(Na Fianna Eireann) Youth wing of the IRA
NICRA	Northern Ireland Civil Rights Association
NIHE	Northern Ireland Housing Executive
O na E	(Oglaigh na Eireann) Dissident Irish republican paramilitary group
OIRA	Official IRA
PIRA	Provisional IRA
PM	Prime Minister
RIC	Royal Irish Constabulary
RIRA	Real IRA
RUC	Royal Ulster Constabulary

SDLP	Social Democratic and Labour Party
SF	Sinn Féin
Taoiseach	Irish Prime Minister
UDA	Ulster Defence Association
UK	United Kingdom
US	United States of America
UUP	Ulster Unionist Party
UVF	Ulster Volunteer Force

1

FOUNDATIONS

Nationalist struggles and geopolitical divisions

On May 17, 2011, Queen Elizabeth visited the Republic of Ireland's Garden of Remembrance, a national memorial and commemorative space for those who died fighting for Irish independence from Britain. The queen, accompanied by the President of the Republic of Ireland, Mary McAleese, laid a wreath on the Garden's altar alongside the national flag of Ireland. Both leaders of state subsequently shared a moment of silence as the Republic's national anthem played.

The Garden, located in Dublin, is a particularly poignant space for many within the Republic. Its location is endowed with a history that has the power to evoke a strong national sentiment – the site where the British government detained many of the leaders of the 1916 Easter Rising before their imprisonment and execution. Fought primarily in Dublin, the 1916 failed insurrection is commonly considered the beginning stage of a violent campaign for independence against British rule in Ireland. Indeed, the Rising, or "the first tremor to shake the British Empire's foundations," became the inciting event of a geopolitical evolution of Ireland that eventually procured a republic on most of the island (Ferguson 2003, 323).

During the Great War (World War I, 1914–1918), the outbreak of an armed insurrection in Ireland initially appalled much of the Irish public. However, as Ireland entered the war under the British flag, many experienced a drastic change of heart after the British state decided to execute many of the leaders of the Rising. Indeed, the verdict of death by firing squad fomented affirmations for independence and an Irish sovereign state among much of the previously reluctant public. As Protestant Irish poet William Butler Yeats pronounced in his famous poem that analyzes the significance of the Rising for Ireland, "All is changed, and changed utterly" (1920).

Throughout the British Isles, this Irish national commemorative garden is imbued with great significance, regardless of whether it is perceived as celebratory or provocative. It reverberates with historical meaning for many, however layered

DOI: 10.4324/9781003141167-1

and contested their personal interpretations may be. While spatial conflict and reconciliation efforts throughout Ireland (i.e., the entire island – both Republic of Ireland and Northern Ireland) are experienced uniquely in different places and by different individuals, commemorative spaces can hold great emblematic power and meaning within a society. Symbolic landscapes can be mechanisms through which efforts to shape national memory and identities are imparted, (re)produced, and challenged (e.g., Hagen and Ostergren 2020; McDowell 2008; Dempsey 2012; Myadar and Davidson 2021; Myadar 2018).

The Garden of Remembrance honors fallen patriots and was strategically designed to provide symbolic validity for the creation of the Irish republic. This symbolic space is also utilized to galvanize political agendas, such as during the queen's visit in 2011. In this way, the site illustrates "the integrated totality of the state's appropriation of the national past . . . within the structure of the state's ideological pursuits: between what is remembered and what is forgotten" (Myadar 2018, 57). Indeed, a state may choose to honor only certain elements of its past in keeping with its political agenda in the present. While this reticent ceremony was simple in nature, this venerative act, situated in a highly symbolic space, signified an opportunity for renegotiating the troubled and volatile geopolitical relation between the two states (Chapter 8 analyzes this symbolic space and geopolitically significant ceremony).

The relationship between the Republic and the United Kingdom in Ireland is often framed through competing claims to space and territorial belonging, power struggles, and socially constructed exclusionary ethnonationalism. Thus, while many celebrated the queen's presence in the Garden, it also provoked a wide variety of protests. For example, Irish republicans opposed permitting the monarch to enter a national sacred space dedicated to those who fought against the British monarchy. Conversely, some unionists and loyalists disapproved of their head of state exhibiting any form of reverence in a place that honored those whom they considered "terrorists against the British state" (Irish Times 2011).

This book investigates these geopolitical challenges and efforts to employ space to redefine divided national allegiances throughout Ireland and within Irish–British relations. To explore the means through which space and landscape become fundamental elements of conflict, nationalism, and peacebuilding efforts in Ireland, this book examines the production and consequences of two distinct political units in Ireland. More specifically, it investigates key geographic themes of bordering, unity, division, and national narratives. This comprehensive thematic approach fosters a greater understanding of representations of states, uneven power structures, competing or shared utilization of space, reconciliation efforts, and renegotiation of divisive national narratives. It also unpacks how perceived "truths," significance, and lived experiences manifest differently across various geographic scales, ranging from the individual to the international.

This book begins and ends in Dublin – the epicenter of the 1916 Rising and the location of the 2011 wreath-laying ceremony. The ceremony illuminates the profound and cumulative means by which space is utilized in negotiations and productions of ethnoterritorial conflicts, efforts for consolidation of state and regional power, conflict transformation, and peacebuilding processes in Ireland. As the Irish Minister of Culture explained in a personal interview:

> The laying of a wreath by Queen Elizabeth in May 2011 in the Garden of Remembrance was a watershed moment for the complex relationship between Ireland and the UK. The shared commemorative ceremony should serve as an example for future collaboration between Ireland and the UK.
>
> *(interview with Minister Deenihan 2015)*

Dublin is the capital of the Republic of Ireland (formally the "Irish Free State"), a state that comprises 26 counties on the island of Ireland. Northern Ireland, established in 1922, consists of the northeastern six counties, has its capital in Belfast, and is a constituent part of the United Kingdom along with England, Scotland, and Wales (see Maps 1.1 and 1.2). Prior to January 31, 2020, both the Republic of Ireland and the United Kingdom (UK) were members of the European Union (EU); now, only the Republic remains. "Brexit" and its subsequent implications for the relationship between Ireland and the UK, especially with regard to movement across the border between the Republic and Northern Ireland, sparked great alarm and trepidation throughout Ireland, generating fears of a renewed threat of increased violence on the island.

Organization of the book

This book draws from over 150 in-depth interviews I conducted between 2011 and 2019 in the Republic of Ireland and Northern Ireland with national and local politicians, teachers and university professors, students, community members, activists, and former paramilitary combatants involved with reconciliation programs with whom I worked or volunteered. The data collected from participants from diverse backgrounds are not interpreted as a homogeneous representation of perceptions in Ireland. Instead, by combining data from a number of participants over time and in various locations, this research highlights some of the key narratives, experiences, and mechanisms of place-based national identities in Ireland. I recorded all interviews with permission (voice recorded or by hand, based on participant's preference) and followed the Human Subjects protocol specified by the Institutional Review Board. My participant observation and critical analysis of the symbolic landscapes discussed in this book draw from ethnographic and landscape studies' methods (e.g., Bryman 2016).

I also conducted extensive archival research and content analysis (Bryman 2016) from a variety of sources (e.g., official recordings of the Government of the Dáil debates; historical and contemporary Irish and British legislation; newspaper reports from *Irish Times, Belfast Telegraph*, and the *BBC*; eyewitness accounts via Irish Bureau of Military History, Bureau of Military History, and the Defense Forces Ireland; Irish diaries written in 1916, 1922, and the 1940s; films, songs, and art displays of 1916, Irish War of Independence/Anglo-Irish War, the Troubles, etc.) accessed online and in physical archives from the Irish American Heritage Center in Chicago, Illinois, and from university archives in the Republic of Ireland.

MAP 1.1 Map of the island of Ireland within the British Isles

The book is organized into eight chapters that examine various and often overlapping themes. Chapters 1, 2, and 3 establish an introductory historical and geographic context that serves as a foundation for the subsequent chapters. Chapters 4–7 focus primarily on the ethnonational sectarian conflict and reconciliation

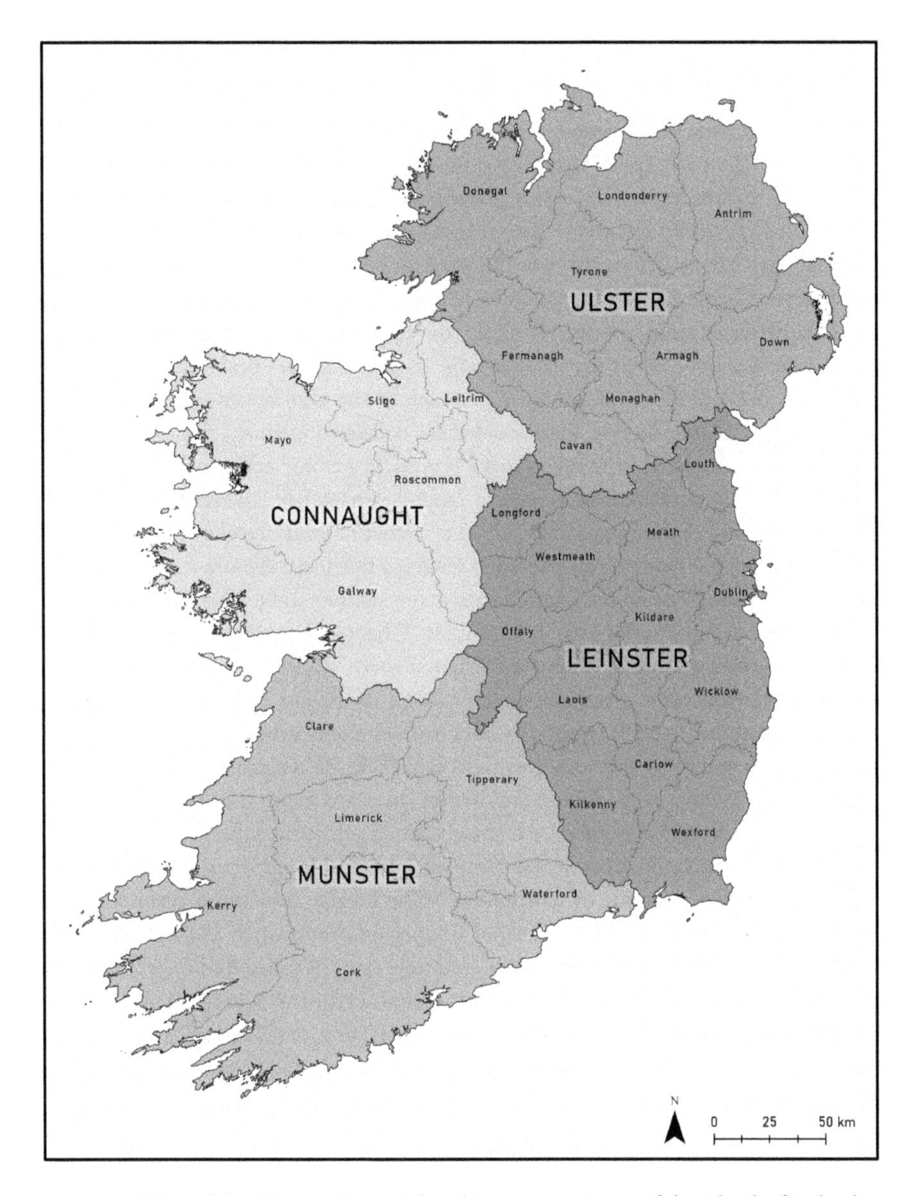

MAP 1.2 Map of the 32 counties and four historic provinces of the island of Ireland

Source: Map by Lauren Andersen.

efforts in Northern Ireland, while Chapter 8 discusses reconciliation efforts between the Republic and the UK.

The remainder of Chapter 1 defines key terms and categories utilized throughout the book and then provides a brief history of national struggles in Ireland. Chapter 2 examines territoriality and nationalism as it recalls the central political and geographic tenets of the insurrection against British rule that transpired during

Easter week, 1916. It contextualizes this relatively small and poorly executed Rising, entangled within the charged milieu of WWI and the subsequent Irish War of Independence/Anglo-Irish War. Chapter 3 explores the role of geography in partition (e.g., bordering processes) and prominent geopolitical events that resulted from the Anglo-Irish Treaty that forged the Irish Free State (with British dominion status) and Northern Ireland. This chapter also chronicles the disputes between republicans regarding this treaty that eventually erupted into the Irish Civil War.

Building on the previous chapters' historical and geopolitical foundations, Chapter 4 examines the spatial aspects of sectarian violence in Northern Ireland through scale, with a multiscalar investigation of the "Troubles" through the 1998 peace agreement. Chapter 5 examines how urban spatial arrangements and strategies both reflect and influence ethnonationalism. More specifically, it explores the foundations and consequences of segregated neighborhoods, social bordering, collective memory, and ethnonational narratives to understand how urban space perpetuates ethnic enclaving in what many have described as apartheid in Northern Ireland. Chapter 6 explores the "meaning" of landscapes and demonstrates how urban and cultural landscapes (including political murals and parades) symbolically mark and reinforce territorial divisions and imaginative boundaries along sectarian lines within many communities in Northern Ireland.

Chapters 7 and 8 offer two case studies. Chapter 7 provides an investigation of a local, grassroots effort to create a "shared space" to forge positive "mix-community" interactions in sectarian-divided areas of Belfast. Chapter 8 returns to the investigation of the Republic's Garden of Remembrance to examine official efforts to renegotiate a fractious past, selective "official" memory, divisive national narratives, and how reconciliation and shared space are (re)produced through intersecting scales of contemporary Irish society. Such efforts may be interpreted as an indication of a renegotiation of the troubled relationship between these states and changing characterizations of nationhood. The book's conclusion returns to the broader context of the island of Ireland to explore events, such as Brexit negotiations (e.g., Northern Ireland Protocol), and how space is utilized to foster efforts to renegotiate perceptions of national identities and territorial belonging.

Before examining Dublin during the 1916 Rising in Chapter 2, the following section introduces the geopolitical underpinnings of the book to contextualize some of the everyday realities of living on an island in which space is highly political.

Geography, geographical understanding of nationalism, and national identities in Ireland

This book introduces complex geographies of peace, conflict, and reconciliation efforts in Ireland from 1916 until the 2021 Brexit transition. It investigates the role space plays in power relations, statecraft, efforts to shape national narratives, and the formation and evolution of place-based identities. It also offers geography as a framework to gain insight into the role that perceptions and utilization of space play in ethnonational territoriality within Ireland – to gain insight into the connections people forge between our surroundings and the different forms of significance we assign to them.

As a discipline, geography

> explores – and promotes critical thinking about – how the world is organized, the environments and patterns that exist on the ground or that humans create in their minds, the interconnections that exist between the physical and human environment, and the nature of places and regions.
>
> *(Murphy 2018, 8)*

Employing a geographic perspective on peace, conflict, and reconciliation in Ireland fosters an examination of *where* and *why* specific things happen in a specific place (i.e., setting of everyday life) and *how* a specific geographic location impacts human and environmental processes. Geographically located events and processes produce distinctive outcomes and perceptions of place. A geographic perspective also focuses on critical inquiries of place-based differences and the social constructions (i.e., ideas or categories widely accepted as factual) of perceptions of people and places.

The intention of this book is to demonstrate how competing and exclusionary claims, perceptions, and employment of space (banal or strategic) produce nationalized conflicts. It also interrogates how these conflicts shape space in Ireland. Additionally, this book explores how space is utilized and reterritorialized in reconciliation efforts, and the production of new civic or non-exclusive national narratives (e.g., Hagen and Ostergren 2020; Kearns, Meredith, and Morrissey 2014). As geographer Brian Graham suggests, the utilization of "representations of landscape and place create manipulated geographies that mesh landscape and memory within the contested arenas of cultural identity and nation-building" (1997, 193). Thus, space is the context, the "scope or reach" of individuals' actions that can range from the local to that of the global, through which geopolitical articulations and banal national relationships are forged, emerge, and are renegotiated across various scales (Flint 2017).

To explore these themes, this book draws on critical geopolitical perspectives. Geopolitics, or a "way of seeing the world" (Flint 2017, 33), examines situated connections among geography, power, politics, diversity, culture(s), and identities (Dodds 2007). Within this approach, the state "operates spatially and manifests in various state-orchestrated projects and strategies" as a geopolitical form of social organization and a spatial expression of power (Moisio et al. 2020). As Foucault (2009) argued, the state operates as an institution that organizes society and disciplines its subjects. Its ability to exert control exists in its capacity to systematize space and regulate its political subjects, even if the state is not the only source of power within a society.

According to Bourdieu (2014), the state represents a monopoly on violence, both symbolic and physical, which exists inside and outside of societal actors via official state acts and delegation of authority. Indeed, while states and state space have changed over time, they continue to constitute an assemblage of legislation, governance, socio-political practices, and national narratives (e.g., McConnell et al. 2017). Based on these assumptions of statecraft and control, critical geopolitical analyses have examined the practices, representations, narratives, sources of power and knowledge, and control of territory (e.g., Dodds 2010; Flint 2017). This includes how borders, a significant constituent of sovereign states, are interpreted as spatial, material, dynamic, economic, societal, and symbolic markers of division and power (e.g., Jones 2016; Brambill and Jones 2020; Papadopoulos and Petridis 2021).

Examining border processes is critical to understanding changing definitions of nationhood and identity politics at the regional (i.e., Northern Ireland) and national scales, as well as within interstate and member state–EU relations (as discussed in Chapter 3). The state's power also exists "in the ability to define identity categories by establishing boundaries between them, and then employing such boundaries as standards of inclusion and exclusion" (Myadar and Rae 2014, 561). Bordering phenomena can reinforce or challenge preconceived notions of national identities (e.g., inclusion or exclusion) as they divide territories and foster categorizations of "us" and "them" that align with territorialist geopolitical imaginings of people and places.

The creation and maintenance of complicated border space by states often contribute to the production of violence (e.g., Dempsey 2020; Dodds 2013; Nevins 2010). Indeed, the territorial nature of border enforcement (e.g., through categorizations of national identities, citizenship, and "foreigner") manufactures the "Other" as different and alien, despite historical connections or commonalities (Dempsey 2022). The socio-political processes of border making or "bordering" emerges through and is maintained by representations in the media, political discourses, language, bureaucratic and cultural practices, textbooks, and socio-spatial conceptualization of human territoriality (e.g., Dempsey and McDowell 2018; Scott 2011). However, this book will demonstrate how national identities' complex and malleable natures cannot always be neatly encompassed within their corresponding sovereign state borders. Additionally, unlike a fixed entity, the physical landscape of the border in Ireland has transformed over time; these significant changes are notably reflected in and influenced by the residents of this international borderland's livelihoods and everyday experiences (Nash and Reid 2013).

The Republic of Ireland gained sovereign statehood in 1949 with the Republic of Ireland Act 1948, but the border in Ireland (i.e., between the Republic and UK's Northern Ireland) remains a topic of contention. While the solidity and porosity of the Irish border have been politically modified over time (e.g., as a result of EU economic integration and the 1998 peace agreement), its presence may continue to be seen as a symbolic and physical reminder of geopolitical divisions, socio-cultural tensions, and contestation over place-based identities. The ire over a border in Ireland, which diminished comparatively as a result of the 1998 peace agreement and subsequent demilitarization of the boundary, reignited during the UK's transition out of the EU. Brexit, or the "British exit," began after the tumultuous 2016 British referendum asked voters, "Should the United Kingdom remain a member of the European Union (EU) or leave the EU?" Brexit supporters framed much of their campaign to leave the EU with anti-immigration narratives and the need to reclaim British sovereignty. To many of Brexit's Leave supporters, the EU represented a "perceived loss of control of their political system" and British sovereign borders (Bachman and Sidaway 2016). However, the spatial and political ramifications of redefining the "permeable" border in Ireland into a hardened one are potentially catastrophic for peace in Ireland.

Brexit challenges the permeability and "invisibility" of the presently inconspicuous Irish border as debates over its future fuel tensions throughout the UK and the Republic of Ireland. During the ensuing UK-EU negotiations, the border in Ireland became a particularly contentious issue when Britain considered the possibility of a "hard exit" strategy. If accepted, it would have relocated the EU's external customs border from the perimeter of the Irish island to the internal border between the Republic of Ireland and Northern Ireland. This insinuates that the border likely would have had to be rearticulated and possibly re-militarized and re-solidified. This would also have ruptured the preestablished rights and movement apportioned to residents on both sides of the UK–Republic of Ireland border. The fear that a "hard" border between the Republic of Ireland and the UK would reignite violence led the British PM and the European Commission to consider an "Irish backstop" to avoid a "hard" border in Ireland. In essence, this approach imperiled the fragile peace fostered by the 1998 peace agreement regarding the border in Ireland.

The increasingly precarious and uncertain conditions for the border infuriated many on the island of Ireland. Some launched satirical critiques of Brexit, such as the Twitter sensation @BorderIrish and the corresponding book, *I am the Border, so I am*. This book depicts the Irish Border as an anthropomorphized line that joins Twitter to voice its criticisms of Brexit negotiations. Through humor, the "border" critically mocks the spatialities of governance and power in post-colonial Ireland as well as the British–Irish–EU strategies of statecraft. As the Irish Border explains:

> I'm a geopolitical line of demarcation between two countries in the EU. I'm also politically contentious, a bit pointless and totally covered in grass . . . and the British government completely forgot about me . . . I started to worry about if I have to become a hard border . . . And then I remember the checkpoints and the soldiers. And the pain. The pain and the mourning . . . And that's how I ended up on Twitter in the middle of this Brexit ruination. It's how I've made myself heard and how, in my own small and insignificant way, I have totally messed up Brexit.
>
> *(7–9, 2019)*

In this way, the book brings to the fore some of the skeptical public's interpretative practices of the geopolitical articulations underpinning "Brexiters'" confident assurances for the future.

Ultimately, the EU and the UK agreed to the Northern Irish Protocol between Northern Ireland (exited the EU) and the Republic (EU member state) to allow rights and mobility to remain primarily as before, with Northern Ireland following many of the former EU regulations. However, since Wales, Scotland, and England are excluded from EU regulations, the regulatory border between Northern Ireland and the rest of the UK sparked consternation among those within Northern

Ireland who do not want their region to be perceived as different from the rest of the UK. Indeed, the presence or modification of state borders can play a significant role in shaping national identities and perceptions of belonging. The establishment of the border in Ireland is explored in Chapter 3, and Brexit proceedings' influence on the border is discussed in the book's conclusion.

Since states are forged through networks of government, civil society, citizenship, nationalism, borders, and governance (Jones 2016), corresponding nationalist geopolitical traditions are highly complex and change over time. Indeed, national identities are fluid and often contradictory intersections of an individual's citizenship, gender, culture, class, ethnicity, and religion (e.g., Massey 2013). Despite their dynamic nature, nationalist place-based identities are integral components of and contribute to the existence of a political state, underpinning ideologies that support its legitimation, validation, as well as preservation of its territorial coalition of places (e.g., Dodds 1993; Koch and Paasi 2016; Myadar 2017). Many states' geopolitical actions reveal efforts to forge or maintain sovereignty by fostering shared visions of national identity and mapping their highly selected official histories onto their bounded territory.

Seminal works on nationalism[1] and national identity, such as Benedict Anderson's (1991) *Imagined Communities*, suggest the state can be a powerful agent that often endeavors to legitimize its power by attempting to foster a sense of shared (and imagined) national identity. Throughout Ireland, there are many examples in which perceptions of belonging are symbolically grounded in nationalist narratives and commemorative spaces. Thus, iconic landmarks, such as the Garden of Remembrance in Dublin, can be politically significant social constructs employed to symbolize a nation of people linked by a common history and territory (e.g., Smith 1995; Billig 1995). How space is utilized, including in conflict or reconciliation processes in Ireland, plays a key role in creating and renegotiating a sense of shared national memory (e.g., Hagen and Ostergren 2020). Thus, these exclusionary and inclusionary geographical imaginings reveal assumptions, interventions, and (re)positionings of bounded space that hold great power and influence in society and the minds of many individuals (e.g., Gregory and Pred 2007).

The discursive politics of nationalism, geopolitics, and identity in Ireland are contested, renegotiated, and dynamic. As the Republic of Ireland continues to become more diverse (e.g., ethnically, religiously, and socially as an outward-facing EU member), the presence of strong national geopolitical traditions, bolstered with key constitutive components, underpins distinct communities across the island. For example, historically, many Irish national narratives highlighted Celtic landscapes, the culture of the West of Ireland, and, eventually, Catholic elements of the island (e.g., Johnson 1994). These traditions and national sacred spaces emphasized perceptions of a shared Irish national identity and appeals for independence from

1 Many scholars of nationalism often describe a spectral range between "civic nationalism" (a more inclusive form that commonly underpins territorial belonging) and "ethnic nationalism" (exclusive, often of ethnic difference) (see Dempsey 2022 for further discussion).

British cultural, religious, and economic oppression (Todd 2018). However, as this book will demonstrate, these highly selective conceptions of nationalist narratives, memory, and identity are regularly altered or challenged. Moreover, they are experienced and expressed differently across disparate socioeconomic backgrounds as well as gendered and racialized lines (e.g., Dempsey 2016, 2022; McCrone and Bechhofer 2015; Koch 2016; Myadar 2017).

Irish national identity is a process often framed by integral components, such as an ethnonational link to the Irish language, landscape, Gaelic history, culture, sports (and the Gaelic Athletic Association), Roman Catholic faith (i.e., spiritually or "culturally"), and non-Britishness (e.g., Conner 2017). Indeed, these highly selective nationalist scripts commonly foster a partisan perception of Irishness, often in contrast to the British "Other" (e.g., Todd 2018). The role of women in Irish nationalism is multifaceted, contested, and often heavily influenced by religious images and domestic expectations (e.g., McDowell 2008; Dowler 2001). Eamon De Valera, a prominent Irish political leader, drew strong connections to the landscape of the west of Ireland. He also adamantly promoted a Gaelic-Catholic narrative of Irish identity, which he integrated into the wording of the 1937 Constitution of Irish Free State (discussed in Chapter 3).

National identities, unpinned by a perceived shared memory, are "subject to relations of power within society, and these power relations work to establish what is generally perceived as the truth within particular domains of knowledge" (Poulter 2018). As national memory often emanates from and is constructed by calculated efforts by the state, "it is usually the official arbitrator of public commemoration and, therefore, of national identity, and as such, it assumes responsibility over planning, maintaining and funding memorial monuments, programs and events" (McDowell 2008). Indeed, many scholars argue that historically, these deliberate, elite-driven, "top-down" efforts to forge an Irish national identity rooted in the Irish landscape proved efficacious for many who identified as Irish (e.g., Graham 1998; Conner 2017).

However, evidence suggests that conceptualizations of Irish national identity within the Republic of Ireland are increasingly diverse (e.g., McGinnity et al. 2018; Central Statistics Office 2016). For example, as Irish-born second-generation Pakistani interviewee from Dublin stated:

> I'm fully Irish. I was born here, learned the Irish language in school and can speak it frequently with any of my mates. I'm a practicing Muslim, I have an Irish passport, I watch Irish television, and Ireland is my home. Yah, I'm Irish.
>
> *(personal interview in Dublin, July 2018)*

National identities in Northern Ireland are diverse and contested, as many continue to self-divide along pervasive partisan, ethnonational lines (e.g., Irish or British) with strongly bounded senses of territory, memory, and belonging (McDowell and Shirlow 2011). For example, while UK citizens are British, they commonly utilize different geographic identifications: English, Scottish, Welsh, respectively, and in Northern Ireland, individuals commonly use either "British," "Irish," or "Northern Irish." Despite the Irish and British states' recognition of the birthright of all the people of Northern Ireland to possess both British and Irish citizenship via the Belfast Agreement (Secretary of State for Northern Ireland 1998), most in Northern Ireland commonly categorize individuals as either Irish (Catholic) or British (Protestant) (e.g., Lloyd and Robinson 2011; Tonge and Gomez 2015; Central Statistical Office 2016).

After the formation of Northern Ireland, the longstanding distrust of Irish Catholics became increasingly conspicuous as the Protestant-controlled devolved regional parliament, known as Stormont, often identified Catholic citizens as "enemy within" (e.g., Barry 2003). This categorization reflected many Protestants' belief that Irish Catholics wanted to abolish Northern Ireland's union with the UK (e.g., Poulter 2018). Like many British within the newly formed Irish Free State, political structures in Northern Ireland rendered Irish residents "second-class citizens," fueling conflict in the region.

In the 1960s, civil rights organizers launched a movement that eventually descended into three decades of violence known as the Troubles, described in Chapter 4. Exclusionary ethnonational categorizations contributed to the tumult and normalized perceptions of divided communities that, for many, persisted after the 1998 peace agreement. Even today, as many schools are segregated along ethnonational lines and often serve as a signifier of one's background, an anecdote for introductions in Northern Ireland often includes questions such as "are you a Protestant or a Catholic?" or "where did you go to primary?"

However, the politics of nomenclature in Northern Ireland underpin multilayered labels that often represent an ethnonational identity expressed simultaneously through religious labels (e.g., Todd 2018). For example, "Catholic" or "Protestant" categorizations may refer to the particular community or neighborhood where an individual resides or the religious background in which one was raised, rather than indicating an individual's religious faith. Indeed, for many, these faith-based terms developed into ethnonational categories with a growing number of people willing to identify with national elements but not the religious components of that classification (e.g., McDowell, Braniff, and Murphy 2017). Some do not employ these labels as they refuse to be classified along religious lines. Others actively work to forge new, more inclusive forms of nationalism that reject oppositional categorizations (see Dempsey 2022). For example, Irish national identities in Northern Ireland can be heterogeneous in nature, as some entwine their "Irishness" with British citizenship. There are also a variety of political ideologies, particularly in Northern Ireland (see Table 1.1).

Irish nationalists and republicans want the unification of the island of Ireland under Irish rule. Prior to 1921, they sought to obtain independence from British rule of Ireland. After the establishment of Northern Ireland, their focus evolved into political struggles over the relationships within what became the Republic and Northern Ireland (Todd 2018). The establishment of a border in Ireland fueled

TABLE 1.1 Common ethnonational categories in Ireland

COMMON ETHNONATIONAL CATEGORIES	
Commonly associated with "Irish Catholicism"	**Commonly associated with "British Protestantism"**
nationalists: desire a united Ireland, for example, SDLP	**unionists:** want to preserve Northern Ireland by maintaining the union with the UK, for example, UUP
republicans: want a united republic in Ireland, traditionally support an armed struggle, for example, Sinn Féin or IRA	**loyalists:** want to defend Northern Ireland, traditionally support an armed struggle, for example, UVF

conflict and provoked nationalists' and republicans' demands for the dissolution of the nascent constituent unit. Within Northern Ireland, most nationalists are affiliated with the Catholic community, regardless of their spiritual beliefs. Often considered less radical than republicans, the categorization can indicate a belief that an independent, united Ireland should be obtained through nonviolent, constitutional means. The Social Democratic Labour Party (SDLP) is nationalists' main constitutional party, especially in Northern Ireland. Republicans consider any form of British rule in Ireland illegitimate. They want to eliminate British control of Northern Ireland and unify Ireland under a republic. Irish republicans are *not* affiliated with the republican party in the US; instead, many in Ireland believe the Republic should be a socialist and unitary state (e.g., Gillespie 2008). The republican classification may describe individuals who support the use of violence to obtain this objective. Their most prominent political party, Sinn Féin, has close links to Irish paramilitary organizations such as the Irish Republican Army (IRA) (e.g., Lally 2020).

While republican historical interpretations tend to describe the demand for Irish independence as continuous (e.g., dating back to the twelfth century and the Anglo-Norman invasion), this over-simplification selectively omits a more complex picture (e.g., Graham 1997; Crowley et al. 2017). For example, some in Ireland are reluctant to recognize or commemorate Irish who fought in the Great War (hereafter WWI) under the British flag, because republicans perceive the enlistment as capitulation to British imperialism (Horne and Madigan 2013). Indeed, Britain and Ireland's long and complicated history, frequently fraught with conflict and violence, spans several centuries and a vast array of political arrangements (e.g., Grayson and McGarry 2016; Dempsey 2018).

Unionists and loyalists want to maintain Northern Ireland's existing link with Britain, oppose Irish rule of Ireland, and venerate their devotion to the British crown (including their contributions during WWI) (Poulter 2018; Braniff, McDowell, and Murphy 2016). Unionists are primarily Protestants with some notable ambiguity within some categorizations, including one's spiritual beliefs, although some Irish (Catholics) identify as unionists (e.g., Central Statistics Office 2016). This term can also indicate a more "mainstream" approach that condemns violence as a means of conserving the British union. Many unionists descended from early seventeenth-century planters – Scottish Presbyterians Protestants, and to a lesser extent, English Anglican Protestants. Unionists' main political parties include the Ulster Unionist Party (UUP) and the more extreme Democratic Unionist Party (DUP). Loyalists support direct action to protect the Protestant ethnonational rule of Ulster from Irish control. The term frequently indicates a more militant form of unionism and is applied when describing paramilitary groups. Loyalist paramilitary groups include the Ulster Volunteer Force (UVF) and the Ulster Defense Association (UDA). Both republican and loyalist paramilitary groups are commonly classified as terrorist organizations (e.g., Moody and Martin 2001), as discussed in Chapter 4.

As many in Northern Ireland compete for control and territory along perceived sectarian lines, the spatial manifestation of these engrained, divided geographies reveals how many have constructed segregated communities and communal perceptions around practices of exclusion (Flint 2004; Graham and Nash 2006; Lloyd and Robinson 2011). Contrasting mental maps of local neighborhoods and partition walls also underpin divided communities and reinforce the perception of ethnonational divisions. Competing national memories, practices, philosophies, and narratives are also spatially mobilized through the tradition of parading, sectarian political murals, and residential segregation, as discussed in Chapters 5 and 6.

However, there is also diversity and discord within each ethnonational group (e.g., McCrone and Bechhofer 2015). For example, there is a variety of nationalism and republican ideologies (e.g., socialist, conservative, primordial "Gaelic," Catholic, etc.) and unionist ideologies (e.g., liberal, civic, ethnic, etc.). Studies also consistently expose disparities within the categorical division of the "binary" communities – that is, among and within the loyalists and unionists (who are often stereotyped as Protestants) as well as the republicans and nationalists (commonly labeled as Catholic) (e.g., Nolan et al. 2014).

There are also prominent generational, cultural, socioeconomic, political, ideological, gender, and religious differences within stereotypically homogeneous communities. Indeed, exclusionary nationalist categorizations and narratives are not universal. For example, a relatively small percentage of young "Protestant" individuals in Northern Ireland prefer to self-identify as "Northern Irish." While this classification may not necessarily signify the presence of an integrated, multi-national community in Northern Ireland, it is evidence of a different form of national self-identification (e.g., Hayes and McAllister 2013; Tonge and Gomez 2015).

Changing definitions of nationhood

Despite the presence of ethnonationalism and partition, there are also examples of more inclusive forms of nationalism. Sports can provide insight into socio-political relations within a nation (e.g., Koch 2017), as "they can offer a window into the world of place, more accurately, a mirror reflecting the process of constructing a place's identity" (Conner 2017). For example, the Ireland National Rugby Union Team (IRFU) is an all-Ireland team that comprises players from the Republic and Northern Ireland. Formed in the 1870s, the team remained "all-Ireland" after partition, unlike national football teams, and offers an example of diverse participants "playing as one island and one Ireland" (O'Driscoll in Williams 2018 film) supported throughout Ireland. As one of the former captains, Willie John McBride explained:

> as an Ulster Protestant, I played for Ireland. In sports, we don't care about politics . . . we just respect the people you're with on the team.
>
> *(cited in Williams 2018 film)*

Since rugby was introduced in Ireland from England, the sport may be considered less "Irish" than, for example, Gaelic football, which is organized by the Gaelic Athletic Association (GAA). The GAA is headquartered in Croke Park in Dublin. Many Irish nationalists and republicans consider the stadium as an Irish national memorial space because British troops attacked and killed several civilian spectators there on "Bloody Sunday" in 1920 (discussed in Chapter 2). Therefore, the 2007 IRFU match against the English national team at Croke Park is significant. The fact that the English team was invited to play on Irish national "hallowed ground" while fans of various ethnonational traditions from the Republic and Northern Ireland attended to support the IRFU demonstrates more cooperative geopolitical relations and a widening field for identity. As the IRFU's captain Brian O'Driscoll suggested:

> Sports bring people together, even though there are two worlds in Ireland. They come together under the watchful eye of the tricolor [Irish flag] . . . and the supporters, when it seems at times there's no commonality for what it means to be Irish, but they stand by one another in the stands and are willing to accept for Irish rugby that we're a whole country . . . Maybe this is modern Ireland, accepting Ireland.
>
> *(cited in Williams 2018 film "Shoulder to Shoulder")*

As the quote reveals, space plays a crucial role in many emerging efforts to renegotiate divided and fractious geopolitical relationships across Ireland and throughout the British Isles. Strategic employment of space can also foster cross-community connections, "teach" reconciliation, and encourage "shared" national ideologies that are not entrenched in "conventional" binary categorizations (e.g., Chapters 7 and 8). For example, during an interview with students who participated in cross-community programming at a "shared" community center in Belfast that is designed to foster friendships that transcend ethnonational lines, one 17-year-old explained:

> Aye, our non-sectarian friendship is rare, especially in north Belfast. But nationalism or religion isn't an issue for my mates because we're all friends

> and we know we're all from here . . . He [pointing at another interviewee from the program] might be British and I'm Irish, but we're both from North Belfast.
>
> *(personal interview, May 2016, cited in Dempsey 2022)*

Particularly since the 1998 Good Friday/Belfast Agreement[2] (hereafter the 1998 peace agreement), some in the Republic and Northern Ireland have worked to diminish the exclusionary and essentialized expressions of identity and belonging. This includes individuals who reject binary ethnonational/religious categories to foster more inclusive forms of civic nationalism. Additionally, despite the conventional image of Northern Ireland as a divided society between Irish-Catholic and British-Protestant, it is important to note the presence of ethnic and religious minorities, including Chinese, Indian, African, Pakistani, and Syrian, as well as "Irish mixed" families (i.e., Irish/Catholic and British/Protestant, discussed in Chapter 5) (e.g., Central Statistics Office 2016; McGinnity et al. 2018). I interviewed many multi-ethnic citizens and/or "mixed" families in both the Republic of Ireland and Northern Ireland. For example, as an individual from a "mixed" family in Belfast, Northern Ireland, explained:

> My mom is Irish, my dad's a Brit. Instead of choosing one side of the family over the other, like many do if they have a mixed marriage in Northern Ireland, my parents put me in a "mixed school." Mine was the Forge Integrated Primary School. Those kinds of primaries aren't that common, but my parents wanted me to have a diverse group of friends and experiences growing up.
>
> *(personal interview, 2017)*

To contextualize the geopolitical understandings of space, borders, and identity (e.g., national, territorial, ethnic) and the spatial production of conflict and reconciliation efforts in Ireland, the remainder of this chapter presents a brief history of the evolving conceptualizations of nationalism and nationalist struggles in Ireland.

A brief history of Irish national struggles

The colonization of Ireland, primarily by English settlers, began in the thirteenth century. Throughout this newly forged colonial space, some of the English assimilated by marrying into Gaelic families and adopting the Irish language and customs, while enclaves of independent Gaelic Irish existed separately from the English throughout Ireland. This new spatial organization of power and residential division

2 The politics of nomenclature in Northern Ireland reflects difficulties embedded within language, labels, place names, and titles that often point to particular ethnonational identities. This can include place names such as Londonderry versus Derry or Northern Ireland versus "the north of Ireland."

reflected hierarchical perceptions of difference. Before long, the presence of foreign settlers as an internal "Other" living in Ireland generated a cultural and linguistic Gaelic Resurgence among many of the native Irish. Efforts to restore their old high-kingship to unite Gaelic Ireland or drive out the English settlers proved unsuccessful.

By the late Middle Ages, the initial English colony, now reduced in size, centered on the area around Dublin, known as the English Pale, a central node of English colonial control of the island. English efforts to rule Ireland continued unevenly, but the Tudor Reconquest of Ireland in the sixteenth and early seventeenth centuries launched a clearly defined period of English rule. The period also witnessed an increased number of political and religious conflicts both in Ireland and throughout Europe. In Ireland, mounting tensions between Irish and English cultures and languages were exacerbated by religious battles between Catholic Ireland and Protestant Anglican England. Control of Ireland and the Irish Sea also played a strategic role in the imperial geopolitical designs of the English as they competed with Catholic Spain and France over colonial claims in North America.

Following a campaign in Ireland that intensified after Henry VIII was named King of Ireland, English efforts to complete the subjugation of Catholic Ireland ultimately succeeded. By the time of Queen Elizabeth Tudor's death in 1603, the old Irish world was gone. The process continued with the flight of the Gaelic earls from Ulster in 1607, which left the northeastern Irish province without local leadership. Subsequently, England established colonial plantations in Ulster (like the Munster Plantation, established in 1586) that drew some English and a larger number of Scottish settlers, primarily Presbyterian, to the area. They were encouraged to move to Ireland to essentially become a garrison community for Britain's own defensive purposes. That immigration would play an important role in forging new geopolitical formations, such as the eventual creation of Northern Ireland. Later, Oliver Cromwell's violent settlement campaign that began in 1649 confiscated all (or, in some cases, a large percentage) of Catholic landowners' property rights on the island. This would ultimately shift Catholic power and affluence in Ireland into Protestant hands, creating a Protestant upper class (Moody and Martin 2001).

After the Williamite victory in the Williamite-Jacobite War in Ireland (1688–1691) at the Battle of the Boyne in 1690, the establishment of the Penal Laws in the seventeenth century severely limited the power of both Catholics and Protestant nonconformists in Ireland. Protestant William of Orange (i.e., King William III of England) deposed Catholic King James II of Scotland, England, and Ireland. To many Protestants in Ireland, his victory was a form of salvation from Catholic rule. As a result, the color orange in Ireland, particularly in the northeast, is associated with him and his supporters, who commonly referred to the monarch as "King Billy" (as discussed in Chapter 6). Under the Penal Laws, Irish resources and territory fell under British control, such that "nowhere else in Europe did a minority oppress a majority in such a fashion . . . and the resultant 'asymmetries of power' proved to be a recipe for intermittent political violence" (Kearns 2007, 9, 12). In reaction, Protestant elite led the first definitive large-scale manifestation of Irish nationalism.

Many Protestant Irish nationalists such as Henry Grattan and revolutionaries such as Theobald Wolfe Tone (who headed the ultimately unsuccessful United Irishmen's Rebellion of 1798 against British rule in Ireland) championed the demand for Irish legislative freedom as an independent republic. The failed rebellion prompted a geopolitical directional change from the government in London, which, against the wishes of Irish republicans who wanted Ireland to become an independent republic, passed the Acts of Union in 1800. These acts dissolved Ireland's parliament in Dublin and legally united Great Britain and Ireland, thereby creating the United Kingdom of Great Britain and Ireland (Smyth 2017).

Over time, many Irish nationalists, including Ireland's "Liberator," Daniel O'Connell, would campaign to repeal the Acts of Union. In fact, during the first half of the nineteenth century, O'Connell emerged as the most prominent champion for Irish rights, demanding the repeal of the Acts and Catholic Emancipation from any restrictive laws against Catholics in Ireland. As a result, O'Connell became one of the most important heroes in Irish history and one of the first Catholics to hold a seat in the British Parliament at Westminster in London. However, he was unsuccessful in his campaign to repeal the Acts, and Ireland did not gain a separate Irish Parliament under the British crown.

Another formidable force in Irish geopolitical agitation, Protestant Charles Stewart Parnell, a member of the Irish Parliamentary Party (IPP), demanded Irish legislative independence and supported legislation for the Government of Ireland Act 1914, commonly known as the Irish Home Rule Bill. The act aimed to create an all-Ireland parliament under the British crown, but not an independent republic. While this bill was unsuccessful, it promised national self-government even if Ireland lacked full independence (e.g., Crowley et al. 2017). Nevertheless, this proposal's existence provoked outrage and great apprehension among many Protestants, particularly unionists and loyalists in northeastern Ireland.

While tensions between Catholics and Protestants existed in Ireland, it is important to not oversimplify unionist opposition to Home Rule as solely a sectarian-driven reaction. For example, compared to other provinces in Ireland, Ulster flourished financially under the Acts of Union, and some unionists feared Home Rule would threaten this prosperity. Others feared potential tariffs and the dismantlement of their welfare pensions for wider redistribution of funds in Ireland. Their strong opposition to Home Rule had profound implications for the future of Anglo-Irish politics and the partition of Ireland (e.g., Callanan 2017). In February 1912, Winston Churchill and David Lloyd George propositioned the British Cabinet that counties in Ulster composed of a Protestant majority should be permitted to leave the proposed Home Rule Bill, and by mid-1913, the state offered a dispensation for any Irish county to opt out of the Home Rule Bill.

Home Rule was suddenly suspended due to the outbreak of WWI in 1914. At this time, Britain possessed massive colonial holdings, including Ireland, India, Asia (South and Southeast), and parts of the African continent. While many Irish, including the IPP, opted to support the British war effort as the war continued, a small group of Irish republicans sought to challenge the British imperial project by attempting to end British rule in Ireland. Consisting primarily of the Irish Volunteers, the Irish Republican Brotherhood (IRB), and James Connolly's Irish Citizen Army (ICA), these rebels staged an armed insurrection on Easter Monday, April 24, 1916, with far-reaching geopolitical ramifications.

This "Easter Rising" occurred throughout the island, but the vast majority of the effort focused on key locations in Dublin. Republicans occupied the city's General Post Office, which served as their garrison headquarters, and was the location from which one of the leaders, Patrick Pearse, publicly read the Proclamation of the Irish Republic. This document was a declaration of Irish sovereignty and war as well as a political and ideological manifesto shared by a minority of supporters in Ireland in 1916 (de Paor 1997). After six days of fighting, however, the rebels surrendered. When the rebellion ended in failure, the government executed 16 of the rebel leaders by firing squad. The executions horrified much of the Irish public and dramatically contributed to strengthening popular support for republicanism throughout Ireland, converting many who previously supported Home Rule into separatists. Growing demands for independence ultimately erupted in the Irish War of Independence/Anglo-Irish War in 1919, which finally ended when Britain and Ireland agreed to call a truce in 1921 (Moody and Martin 2001).

Following lengthy negotiations, the signing of the Anglo-Irish Treaty created the Irish Free State within the Dominion of the British Commonwealth. It was not a republic but embodied what Irish revolutionary Michael Collins referred to as the "freedom to achieve freedom" (e.g., Heintz 2009). Twenty-six of the 32 counties in Ireland would accept the conditions of the Irish Free State in hopes of ultimately achieving an independent republic. The remaining six counties in Ulster, which maintained a British majority since the Ulster plantations began in the seventeenth century, were dominated politically by unionist opposition to an independent Ireland and elected to leave the Irish Free State (Callanan 2017). In 1922, those counties formed the constituent unit of Northern Ireland within the United Kingdom (see Map 1.3).

However, the partition of Ireland is not the only tumultuous rift that transpired on the island during the 1920s. Indeed, after the conclusion of the Irish War of Independence/Anglo-Irish War, debates regarding the future of Ireland began to divide republicans. Against this politically combative backdrop, and after six months of political wrangling, Britain and representatives of the republican Sinn Féin delegation signed the controversial aforementioned Anglo-Irish Treaty. Contestation regarding acceptance or rejection of this treaty precipitously divided republicans. Indeed, the Anglo-Irish Treaty, which solidified the loss of six of nine counties in Ulster and dominion status in the remaining part of Ireland, fueled discord within the Free State and for many who identified as Irish within the newly created Northern Ireland. These tensions erupted in 1922 as the Irish Civil War between pro-treaty and anti-treaty forces. The pro-treaty side ultimately prevailed with the support of British supplies and ammunition.

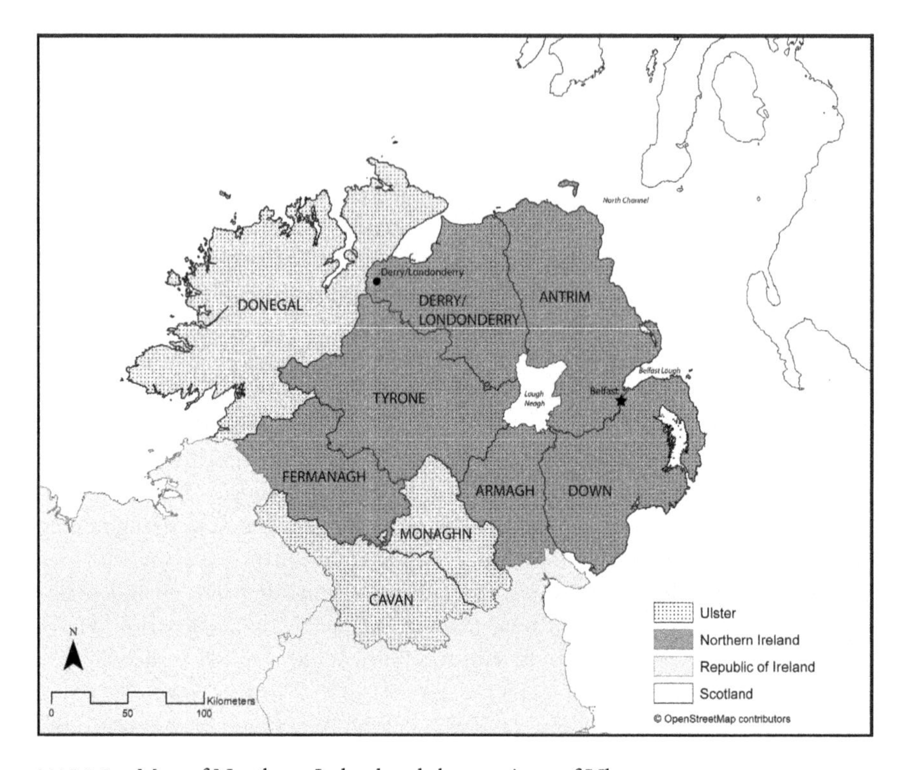

MAP 1.3 Map of Northern Ireland and the province of Ulster

The ratification of the Republic of Ireland Act 1948 officially severed the connection between the British Commonwealth in 1949 and transformed the Free State into an independent republic. However, the strained relationship between the two states became, at times, tumultuous; the era commonly known as the "Troubles" in Northern Ireland spanned the 1960s to 1998. At its core, the Troubles was a struggle over power and space – how power should be distributed and exercised, by whom and at what scale. During the Troubles, the border region in Ireland became increasingly militarized, including the presence of army checkpoints near customs posts. Military presence and sectarian violence created a "no man's land" throughout this border space, effectively creating a "bandit country."

The border area was an IRA stronghold where their paramilitary forces launched a violent campaign targeting the RUC, loyalist paramilitaries, and Protestant civilians in the area. The border violence took an enormous toll on surrounding towns' economies, fomented illegal smuggling and arms deals, and drove many civilians from the disputed area (e.g., Patterson 2013). The border also represents various contemporary layered and contested territorial claims on the island. For example, much of unionists' and loyalists' anxiety as minorities on the island was exacerbated by the Republic of Ireland's declaration that its national territory rightfully comprised "the whole island of Ireland."

This territorial claim to the whole island was later revoked as part of the 1998 peace agreement, as were the military checkpoints erected along the border between the two states during the Troubles (by 2005). However, during the Troubles, the greatest concentration of violence occurred in Belfast and other cities where sectarian or "opposing" enclaved communities closely neighbored one another. The Irish border became a physical and symbolic manifestation of Irish–British geopolitical divisions, inter- and intra-Irish ethnonational divisions, and powerful exclusionary nationalist narratives competing for territorial claims for a homeland in Ireland (discussed in Chapters 4–6).

After years of violence and segregation, in 1998, a diverse array of regional political representatives within Northern Ireland, the Irish and British states, and EU leadership supported the 1998 Belfast/Good Friday Peace Agreement in Northern Ireland. The peace agreement and EU integration helped alter the Irish border's impact by transforming it from a line of dispute or "division" to a "place of meeting." Before Brexit negotiations (discussed in the book's conclusion), the Irish border shed much of its functional significance, becoming more permeable and invisible with time and cross-border cooperation.

The subsequent demilitarized "soft" border in Ireland and efforts made in symbolically significant space, such as the wreath ceremony in the Garden of Remembrance, are intentionally designed to (re)negotiate new geopolitical relationships, shared spaces, and non-discriminative national narratives between the two states. Additionally, the subsequent St. Andrews Agreement in 2006 helped address unresolved power-sharing concerns, further underpinning efforts to foster a fragile peace amidst a legacy of violence in Northern Ireland from which its people continue to negotiate and emerge. However, as discussed in the book's conclusion, such ventures experienced significant challenges as "Brexit" negotiations threatened to return to a "hard" Irish border (since it no longer territorially delineates a boundary jointly shared by EU members). In order to investigate these geopolitical challenges, the following chapter provides insight into the historical alliances and ruptures between Ireland and the British Empire. More specifically, it examines the rise of Irish nationalism in the early twentieth century by focusing on the geographic concentration of violence that transpired during the 1916 Rising and the Irish War of Independence/Anglo-Irish War.

References

Anderson, Benedict. 1991. *Reflections on the Origin and Spread of Nationalism*. London, New York: Verso.

Bachmann, Veit, and James D. Sidaway. 2016. "Brexit Geopolitics." *Geoforum* 77: 47–50.

Barry, John. 2003. "National Identities, Historical Narratives and Patron States in Northern Ireland'." In *Political Loyalty and the Nation-State*, 189–205. Oxfordshire: Taylor and Francis.

Billig, Michael. 1995. *Banal Nationalism*. London: Sage.

@BorderIrish 2019. *I am the Border, so I am*. New York, NY: HarperCollins.

Bourdieu, Pierre. 2014. *On the State: Lectures at the College de France 1989–1992*. Edited by P. Champagne, R. Lenoir, F. Poupeau, and M. Riviere. Malden, MA: Polity.

Brambilla, Chiara, and Reece Jones. 2020. "Rethinking Borders, Violence, and Conflict: From Sovereign Power to Borderscapes as Sites of Struggles." *Environment and Planning D: Society and Space* 38 (2): 287–305.
Braniff, Máire, Sara McDowell, and Joanne Murphy. 2016. "Editorial Introduction." *Irish Political Studies* 31 (1): 1–3. https://doi.org/10.1080/07907184.2015.1126923.
Bryman, Alan. 2016. *Social Research Methods*. Oxford: Oxford University Press.
Callanan, Frank. 2017. *The Home Rule Crisis*. Edited by John Crowley, Donal Ó Drisceoil, and Mike Murphy. Atlas of the Irish Revolution. Cork: Cork University Press.
Central Statistics Office. 2016. "Irish Census Report." https://www.cso.ie/en/media/csoie/newsevents/documents/pressreleases/2017/prCensussummarypart1.pdf
Conner, Neil. 2017. "Sports and the Social Integration of Migrants: Gaelic Football, Rugby Football, and Association Football in South Dublin." In *Critical Geographies of Sport*, 190–206. Oxfordshire: Routledge.
Crowley, John, Donal Ó Drisceoil, Michael Murphy, John Borgonovo, and Nick Hogan. 2017. *Atlas of the Irish Revolution*. New York University Press.
De Paor, Liam. 1997. *On the Easter Proclamation and Other Declarations*. Dublin: Four Courts Press Ltd.
Dempsey, Kara E. 2012. " 'Galicia's Hurricane': Actor Networks and Iconic Constructions." *Geographical Review* 102 (1): 93–110.
———. 2016. "Competing Claims and Nationalist Narratives: A City/State Debate in a Globalising World." *Tijdschrift Voor Economische En Sociale Geografie* 107 (1): 33–47.
———. 2018. "Creating a Place for the Nation in Dublin." *The City as Power: Urban Space, Place, and National Identity* 27–40.
———. 2020. "Spaces of Violence: A Typology of the Political Geography of Violence Against Migrants Seeking Asylum in the EU." *Political Geography*, 1–10.
———. 2022. "Fostering Grassroots Civic Nationalism in an Ethno-Nationally Divided Community in Northern Ireland." *Geopolitics*, 1–17. https://doi.org/10.1080/14650045.2020.1727449.
Dempsey, Kara E., and Sara McDowell. 2018. "Disaster Depictions and Geopolitical Representations in Europe's Migration 'Crisis.'" *Geoforum* 98: 153–160.
Dodds, Klaus. 1993. "Geography, Identity and the Creation of the Argentine State." *Bulletin of Latin American Research* 12 (3): 311–31.
———. 2007. *Geopolitics: A Very Short Introduction*. Oxford: Oxford University Press.
———. 2010. "Flag Planting and Finger Pointing: The Law of the Sea, the Arctic and the Political Geographies of the Outer Continental Shelf." *Political Geography* 29 (2): 63–73.
———. 2013. " 'I'm Still Not Crossing That': Borders, Dispossession, and Sovereignty in *Frozen River* (2008)." *Geopolitics* 18 (3): 560–83. https://doi.org/10.1080/14650045.2012.749243.
Dowler, Lorriane. 2001. "No Man's Land: 'Gender and the Geopolitics of Mobility in West Belfast, Northern Ireland." *Geopolitics* 6, no. 3: 158–76.
Ferguson, Niall. 2003. *Empire: How Britain Made the Modern World*. City of Westminster, London: Penguin.
Flint, Colin. 2004. *Spaces of Hate: Geographies of Discrimination and Intolerance in the USA*. East Sussex: Psychology Press.
———. 2017. *Introduction to Geopolitics*. 3rd ed. Oxfordshire: Routledge.
Foucault, Michel. 2009. *Security, Territory, Population: Lectures at the Collège de France, 1977–1978*. Edited by Michel Senellart, François Ewald, and Alessandro Fontana. Translated by Graham Burchell. 1. Picador ed. Lectures at the Collège de France. New York, NY: Picador.

Gillespie, Gordon. 2008. *Years of Darkness: The Troubles Remembered*. Dublin: Gill & Macmillan.

Graham, Brian. 1997. *In Search of Ireland: A Cultural Geography*. East Sussex: Psychology Press.

———. 1998. "Contested Images of Place among Protestants in Northern Ireland." *Political Geography* 17 (2): 129–44.

Graham, Brian, and Catherine Nash. 2006. "A Shared Future: Territoriality, Pluralism and Public Policy in Northern Ireland." *Political Geography* 25 (3): 253–78. https://doi.org/10.1016/j.polgeo.2005.12.006.

Grayson, Richard S., and Fearghal McGarry. 2016. *Remembering 1916: The Easter Rising, the Somme and the Politics of Memory in Ireland*. Cambridge: Cambridge University Press.

Gregory, Derek, and Allan Pred. 2007. *Violent Geographies: Fear, Terror, and Political Violence*. Oxfordshire: Taylor & Francis.

Hagen, Joshua, and Robert C. Ostergren. 2020. *Building Nazi Germany: Place, Space, Architecture, and Ideology*. Lanham, MD: Rowman & Littlefield.

Hayes, Bernadette, and Ian McAllister. 2013. *Conflict to Peace: Politics and Society in Northern Ireland over Half a Century*. Manchester: Manchester University Press.

Heintz, Matthew. 2009. "The Freedom to Achieve Freedom: Negotiating the Anglo-Irish Treaty." *Intersections* 10 (1): 431–51.

Horne, John, and Edward Madigan. 2013. "Towards Commemoration: Ireland in War and Revolution, 1912–1923." Dublin: Royal Irish Acad.

Irish Times. 2011. "McAleese Hails 'Extraordinary Moment.'". www.irishtimes.com/news/mcaleese-hails-extraordinary-moment-1.876601.

Johnson, Nuala C. 1994. "Sculpting Heroic Histories: Celebrating the Centenary of the 1798 Rebellion in Ireland." *Transactions of the Institute of British Geographers*: 78–93.

Jones, Reece. 2016. *Violent Borders: Refugees and the Right to Move*. Brooklyn, NY: Verso Books.

Kearns, Gerry. 2007. "Bare Life, Political Violence, and the Territorial Structure of Britain and Ireland." In *Violent Geographies: Fear, Terror and Political Violence*, edited by Derek Gregory and Allan Pred, 7–36. New York, NY: Routledge.

Kearns, Gerry, David Meredith, and John Morrissey. 2014. *Spatial Justice and the Irish Crisis*. Dublin: Royal Irish Academy.

Koch, Natalie. 2016. "Is Nationalism Just for Nationals? Civic Nationalism for Noncitizens and Celebrating National Day in Qatar and the UAE." *Political Geography* 54: 43–53.

———. 2017. *Critical Geographies of Sport: Space, Power and Sport in Global Perspective*. 1st ed. Oxfordshire: Taylor & Francis.

Koch, Natalie, and Anssi Paasi. 2016. "Banal Nationalism 20 Years on: Re-Thinking, Re-Formulating and Re-Contextualizing the Concept." *Political Geography* 54: 1–6.

Lally, Connor. 2020. "Untangling the Links Between Sinn Féin and the IRA." *Irish Times*, February 21, 2020.

Lloyd, Katrina, and Gillian Robinson. 2011. "Intimate Mixing – Bridging the Gap? Catholic-Protestant Relationships in Northern Ireland." *Ethnic and Racial Studies* 34 (12): 2134–52.

Massey, Doreen. 2013. *Space, Place and Gender*. Hoboken, NJ: John Wiley & Sons.

McConnell, Fiona, Merje Kuus, Alex Jeffrey, Heaven Crawley, Nick Vaughan-Williams, and Adrian Smith. 2017. "Interventions on Europe's Political Futures." *Political Geography* 60: 261–71.

McCrone, David, and Frank Bechhofer. 2015. *Understanding National Identity*. Cambridge: Cambridge University Press.

McDowell, Sara. 2008. "Commemorating Dead 'Men': Gendering the Past and Present in Post-Conflict Northern Ireland." *Gender, Place & Culture* 15 (4): 335–54. https://doi.org/10.1080/09663690802155546.

McDowell, Sara, Máire Braniff, and Joanne Murphy. 2017. "Zero-Sum Politics in Contested Spaces: The Unintended Consequences of Legislative Peacebuilding in Northern Ireland." *Political Geography* 61: 193–202.

McDowell, Sara, and Peter Shirlow. 2011. "Geographies of Conflict and Post-Conflict in Northern Ireland: Conflict in Northern Ireland." *Geography Compass* 5 (9): 700–9. https://doi.org/10.1111/j.1749-8198.2011.00444.x.

McGinnity, Frances, Eamonn Fahey, Emma Quinn, Samantha Arnold, Bertrand Maitre, and Philip O'Connell. 2018. *Monitoring Report on Integration 2018. ESRI Report*. Dublin: Department of Justice and Equality.

Moisio, Sami, Andrew EG Jonas, Natalie Koch, Christopher Lizotte, and Juho Luukkonen. 2020. "Changing Geographies of the State: Themes, Challenges and Futures." In *Handbook on the Changing Geographies of the State*. Cheltenham: Edward Elgar Publishing.

Moody, Theodore William, and Francis X. Martin. 2001. *The Course of Irish History*. New York, NY: Roberts Rinehart Pub.

Murphy, Alexander B. 2018. *Geography: Why It Matters*. Cambridge: Polity Press.

Myadar, Orhon. 2017. "The Rebirth of Chinggis Khaan: State Appropriation of Chinggis Khaan in Post-Socialist Mongolia." *Nationalities Papers* 45 (5): 840–55.

———. 2018. "The City, Memory, and Ideology in Ulaanbaatar." *The City as Power: Urban Space, Place, and National Identity* 57.

Myadar, Orhon, and Ronald A. Davidson. 2021. "Remembering the 'Comfort Women': Geographies of Displacement, Violence and Memory in the Asia-Pacific and Beyond." *Gender, Place & Culture* 28 (3): 347–69.

Myadar, Orhon, and James Deshaw Rae. 2014. "Territorializing National Identity in Post-Socialist Mongolia: Purity, Authenticity, and Chinggis Khaan." *Eurasian Geography and Economics* 55 (5): 560–77.

Nash, Catherine, and Bryonie Reid. 2013. *Partitioned Lives: The Irish Borderlands*. 1st ed. Oxfordshire: Routledge.

Nevins, Joseph. 2010. *Operation Gatekeeper and beyond: The War on" Illegals" and the Remaking of the US – Mexico Boundary*. New York, NY: Routledge.

Nolan, Paul, Dominic Bryan, Clare Dwyer, Katy Hayward, Katy Radford, and Peter Shirlow. 2014. *The Flag Dispute: Anatomy of a Protest*. Belfast: Institute of Irish Studies, Queen's University.

Papadopoulos, Alex G., and Triantafyllos G. Petridis. 2021. *Hellenic Statecraft and the Geopolitics of Difference*. Oxfordshire: Routledge.

Patterson, Henry. 2013. *Ireland's Violent Frontier: The Border and Anglo-Irish Relations during the Troubles*. New York, NY: Springer.

Poulter, John. 2018. "The Discursive Reconstruction of Memory and National Identity: The Anti-War Memorial the Island of Ireland Peace Park." *Memory Studies* 11 (2): 191–208.

Secretary of State for Northern Ireland. 1998. "The Belfast Agreement: An Agreement Reached at the Multi-Party talks on Northern Ireland." Northern Ireland Office. https://assets.publishing.service.gov.uk/government/uploads/system/uploads/attachment_data/file/1034123/The_Belfast_Agreement_An_Agreement_Reached_at_the_Multi-Party_Talks_on_Northern_Ireland.pdf

Scott, James Wesley. 2011. "Borders, Border Studies and EU Enlargement." In *The Ashgate Research Companion to Border Studies*, edited by Doris Wastl-Walter, 123–42. Oxfordshire: Routledge.

Smith, Anthony. 1995. *Nations and Nationalism in a Global Era*. Oxford: Polity Press.
Smyth, William. 2017. "Nineteenth-Century Ireland: Transformed Contexts and Class Structures." In *Atlas of the Irish Revolution*, edited by John Crowley, Donal Ó Drisceoil, and Mike Murphy, 4–55. New York, NY: New York University Press.
Todd, Jennifer. 2018. *Identity Change after Conflict: Ethnicity, Boundaries, and Belonging in the Two Irelands*. London: Palgrave Macmillan.
Tonge, Jonathan, and Raul Gomez. 2015. "Shared Identity and the End of Conflict? How Far Has a Common Sense of 'Northern Irishness' Replaced British or Irish Allegiances since the 1998 Good Friday Agreement?" *Irish Political Studies* 30 (2): 276–98.
Williams, Isobel. 2018. *Shoulder to Shoulder*. Film. BT Sports Films.
Yeats, William Butler. 1920. *Easter, 1916*, Vol. 25, 69. New York, NY: The Dial.

2

CONTESTED TERRITORIALITIES

The 1916 Rising and the Irish War of Independence

This chapter returns to Dublin, during the 1916 Rising, to explore key geographic themes of employment of space, socio-political bordering practices, and nationalist narratives. It begins by examining this relatively small and poorly executed insurrection within the larger context of WWI and the suspended 1914 Home Rule Bill. This chapter then explores subsequent geopolitical events that contributed to the growth of Irish Republican Army (IRA), the outbreak of the Irish War of Independence/Anglo-Irish War (1919–1921), and the uneven geographies of this armed conflict. The scope of this chapter extends beyond these concentrated eruptions of violence to include the Anglo-Irish Treaty and partition as part of a larger transformative period that repositioned geopolitical relations between Ireland and the British state.

The growing support for republicanism during this time is portrayed in the 2006 film, *The Wind that Shakes the Barley*. The protagonist, Damien, played by Cillian Murphy, alters his plans to move to London for medical school and instead joins the IRA. In the film, when talking to a British soldier, he declares:

> Your presence here is a crime. You tell me, what I am supposed to do? Turn the other cheek for another 700 years? Is that it? . . .

(Later in the film, he suggests):

> Here's what we'll do. We send a message to the British Cabinet that will echo and reverberate around the world. If they bring their savagery over here, we will meet it with a savagery of our own.

DOI: 10.4324/9781003141167-2

The transformation of Damien's perception of London (and Britain) as a desirable place (in his case, for an education) to a criminal entity that should be removed by force reveals how many in Ireland began to reconceptualize British control of Ireland during this time. However, this change was not widely shared in Ireland until after the 1916 Rising. This chapter traces this transformation and subsequent geopolitical repositioned forms of governance and nationalized identities.

Contextualizing the Rising

The Rising altered the course of Irish–British geopolitical relations as it represented the beginning of armed action and a period of violent turmoil in Ireland not witnessed since Wolfe Tone's 1798 rebellion (Callanan 2017). Additionally, the demand for an independent republic ended the prospect of a peaceful settlement to Home Rule negotiations that dominated political discussions prior to WWI (McGarry 2011). During the Rising, Dublin became a stage on which contentious and divided constructions of nationalism, identity, and territoriality played out. For example, many republicans feared the erosion of "Irishness" (e.g., Gaelic culture, language, and traditions) under British imperialism and the "Anglicization" of Ireland. As former Taoiseach, or Prime Minister of the Republic of Ireland, Garret FitzGerald explained, one of the main motivations for rebels' involvement in the Rising was the belief that:

> Without a dramatic gesture of this kind, the sense of national identity that had survived all the hazards of the centuries would flicker out ignominiously within their lifetime.
>
> *(cited in Ferriter 2015)*

Also, many Irish Catholics who suffered partisan bigotry felt excluded from the Act of Union (1800). That geopolitical alienation, paired with the fear of WWI conscription and Home Rule debates, caused many to support Irish nationalist movements (McCarthy 2012).

Prior to the Rising, Irish republican James Connolly unfurled a banner on Liberty Hall[1] that proclaimed: "We serve neither King nor Kaiser, but Ireland" (Skinnider 1917). This geopolitical statement not only demonstrates their dedication to the Irish nation but also reveals republican refusal to accept British rule or political subordination to Germany. While many rebels fought to save the soul of the Irish nation,[2] others interpreted the Rising as a geopolitical-religious act to reclaim Ireland for "its people" (McCarthy 2012). This sentiment underpinned

1 Liberty Hall, located along the River Liffey in Dublin, served as the headquarters of socialist-republican Irish Citizen Army, a soup kitchen (organized by Constance Markievicz and Maud Gonne during the 1913 Dublin Lock-out), and the meeting point where rebels assembled to begin the 1916 Rising.

2 Many rebels were inspired by other European cultural and linguistic nationalist revivals as part of the "Generation of 1914" (see Augusteijn 2010).

hostility toward the British state and a shared conviction to protect their distinct Irish homeland and culture. For example, the decision to attack on Easter Sunday provided the rebels with an opportunity to catch government forces unprepared and the potential for symbolic resonance with themes of redemption and rebirth (e.g., Augusteijn 2010).

The geopolitical nature of the symbolic connection among Irish nationalism, rebirth, and the Rising is also prominent in many memorial spaces throughout Ireland (discussed in Chapter 8). Indeed, the rebellion initiated an evolution of nationhood, forged by the rebels' sacrifice and propaganda, that united much of Irish nationalism under the banner of republicanism. This movement continued to gain momentum, eventually leading to the outbreak of the Irish War of Independence/Anglo-Irish War (McGarry 2010).

Internal divisions and paramilitaries in Ireland

Not everyone in Ireland wanted independence. The subsequent tumult exposed competing power struggles underpinned by contested and partitioned geopolitical conceptions of belonging in Ireland. For example, the previously suspended Home Rule bill, which would have created a devolved parliament in Dublin, antagonized many Protestants, particularly those concentrated in northeastern Ulster. This geographic concentration of unionists played a prominent role in the partition of the island. Unionists, many of whom were Protestants, wanted to maintain Ireland's current geopolitical status within the UK and opposed relinquishing power to "Catholic-dominated" Dublin. Discussions of Home Rule fueled a growing form of Protestantism among many unionists, who feared becoming a minority on the island under "Catholic rule," as well as the possible suppression of their economically prosperous linen and shipbuilding industries in the northeast. It also fostered a communal sense of national identity and territorial belonging among Protestants, particularly in Ulster (McGarry 2010). The paramilitary Ulster Volunteer Force (UVF) was formed in 1912 to challenge the proposed Home Rule bill. Over time, the UVF became increasingly radicalized as its members swore an oath to defy any legislation for Irish independence if granted by the British Parliament (Mansergh 2017).

The creation of a loyalist UVF prompted Irish nationalists/republicans to form the Irish Volunteers in 1913 to counter the growing presence of the UVF. That same year, the Irish Citizen Army (ICA), a smaller socialist militia led by James Connolly, was formed to defend workers' rights during the 1913 Dublin Lock-out strike.[3] The ICA and Irish Volunteers sometimes cooperated with the secret militant revolutionary organization, the Irish Republican Brotherhood (IRB), led by Tom Clarke. Formed in 1858, the IRB worked to forge an independent Ireland by force. Despite the illegality of arms possession in Ireland, by 1914, the Irish

3 For a background on the 1913 major industrial dispute, Conor Larkin, and the ICA, see Crowley et al. 2017.

Volunteers and IRB collected almost 50,000 arms. In addition, republican Irish Volunteers (and the UVF) had their own uniforms, weapons, and publicly practiced military formations (O'Halpin 2001).

The outbreak of WWI divided paramilitary/revolutionary organizations in Ireland. For example, members of the Irish Volunteers split when John Redmond, the leader of the Irish Parliamentary Party (IPP), encouraged the organization to join the war effort to demonstrate their cooperation with the British crown. He hoped their involvement would advance the IPP's campaign for Home Rule. The majority of Irish Volunteers, known as the Redmonite "National" Volunteers thereafter, joined other Irish citizens and participated in WWI (Johnson 2003). However, while the majority of Irish Volunteers (close to 170,000) answered his call, his appeal had significant political repercussions, as it divided the organization's membership. Estimates suggest a notably smaller portion of the organization (between 12,000 and 15,000 of their members), split away from the Redmonite "National" Volunteers and remained in Ireland. Under the leadership of Eoin MacNeill, this faction maintained the name Irish Volunteers (or Óglaigh na hÉireann). Then, close to 2,000 of MacNeill's anti-conscription Irish Volunteers were further radicalized as part of the IRB. This organization would be a central strategic force during the Rising. Many of those executed after the Rising were members of the IRB, including the provisional president of the declared republic, Patrick (Padraig) Pearse (White 2017).

The presence and armament of the UVF, Irish Volunteers, IRB, and ICA represented the increasing militarization and division of geopolitical identities in Irish society. Despite their competing geopolitical objectives for Ireland, the presence of these paramilitary organizations also signified the beginning of a local, territorialized undermining of British rule on the island. However, the British state was hesitant to react. It did not want to provoke the Irish public by attacking forces it believed too insignificant to effectively challenge British rule in Ireland. Additionally, in order to disarm the Irish Volunteers, IRB, or ICA, the state would also have to do the same to the loyalist UVF. It believed such intervention measures would result in bloodshed, further radicalization of Irish nationalism/republicanism, and decreased Irish participation in WWI. Instead, the state focused on the war efforts and its propaganda campaign to inspire Irish to enlist in WWI (Johnson 2003; McCarthy 2012).

British propaganda did little to quell Irish nationalist/republicans' determination for action. By January 1916, the IRB, ICA, and members of the Irish Volunteers, with the support of Cumann na mBan (Irish Women's Council), agreed to jointly organize an armed insurrection to end British rule in Ireland. These organizations also vowed to establish an independent Irish republic before the conclusion of WWI. They believed that 1916 was an opportune moment to "take up the torch of revolt" as the British state was distracted with its war efforts. The organizers agreed that Easter Sunday 1916 would be the start date of their rebellion for religious and strategic purposes.

Missed opportunities and confusion

The groups initiated plans to garner support for their armed struggle from the German government. With much of Britain distracted by war, they hoped their rebellion would be successful with the help of the German state. A Protestant Irish republican, Roger Casement, who co-founded the Irish Volunteers, succeeded in convincing the German state to supply the Irish rebels with 25,000 rifles and 1 million rounds. Germany believed it could benefit from a conflict in Ireland if it divided British attention and undermined their focus on the Western Front (aan de Wiel 2017). However, due to a miscommunication, none of the rebels met the German ship carrying guns and munitions when it arrived in Ireland. Then, when discovered instead by the authorities, the German captain scuttled the ship, thereby terminating the rebels' potential supply line. Despite German and Irish rebels' efforts, the British state continued to underestimate the fortitude of republicanism in Ireland. For example, Dublin Castle, the seat of British power in Ireland, received intelligence on April 18 regarding the large German shipment of arms destined for the insurrection planned for Easter Sunday. However, once the authorities arrested Roger Casement and saw the German ship scuttled off the coast, they believed the rebels would abandon their insurrection plans.

The loss of the German arms indeed proved devastating for the rebels. Debates regarding the full significance of the lost supply shipment from Germany divided them. For example, Eoin MacNeill refused to encourage his Irish Volunteers to act without foreign aid. In contrast, the IRB and the ICA wanted to proceed. When MacNeill discovered the IRB and ICA's plans for the secret armed insurrection on Easter Saturday, he ordered his Irish Volunteers to disregard any commands originating from Patrick Pearse of the IRB. MacNeil also posted placards stating, "All Easter maneuvers canceled by Eoin MacNeill" (eyewitness account, Walsh 1949, 11). His refusal to proceed infuriated bellicose rebels who later named MacNeill's men the "wait-a-whiles" (fan go fóills).

The conflicting accounts of "proceed" and "disregard" confused the rebels. For example, as one of the participants in the Rising, Robert Holland, explained:

> I knew by Sunday morning's paper that the general mobilization was cancelled, but a number of us were in doubt about it being permanent, as we expected that a leakage of our intentions would get out and the press would be against us.
>
> (1949, *Witness, Irish Bureau of Military History*)

Despite armament setbacks and confusion, the IRB and ICA were determined to forge ahead with the rebellion on Easter Monday – one day later than originally planned. They contrived a decoy maneuver to deceive MacNeill and British forces in Ireland before their attack. However, this only further confused would-be participants, and fewer joined the Rising than the groups anticipated.

Dublin as conflict space

The 6-day rebellion began on Easter Monday, April 24, and ended on Saturday, April 29, 1916 (for details, see Crowley et al. 2017). Originally intended to be an island-wide insurrection, the fighting was primarily concentrated in Dublin, with some 1,200 Irish separatists facing over 20,000 British troops resulting in nearly 500 fatalities (McGarry 2011). The IRB served as the main coordinating force for the Rising, accompanied by Irish Volunteers and the ICA. Cumann na Mban (Irish Women's Council) and Na Fianna Éireann (republican youth organization)[4] supported their efforts.

When the rebels began their siege of Dublin, the city was relatively quiet, as only a few hundred British troops remained in the city for Easter and much of the public left to watch horse racing in Fairycastle, outside of Dublin.[5] The rebels occupied certain locations around the city center, some of which, such as a bakery, were of little tactical significance (Yeates 2012). However, their strategy to garrison their forces within substantial structures throughout Dublin in anticipation of heavily armed British reinforcements proved effective for their poorly equipped and soon-outnumbered rebellion.

One such location was the General Post Office (GPO) on Sackville Street (renamed O'Connell Street in 1924). Selected for its imposing construction and central location on a main thoroughfare, it served as a headquarters and key administrative location throughout the rebellion. It was also the location where Patrick Pearse publicly read the Proclamation of the Irish Republic and broadcast promotional reports of the rebels' success throughout the week. The GPO was also the location from which rebels raised the Irish tricolor flag and the green Irish Republic flag. Additionally, the ICA raised the Irish socialist flag, the Starry Plough, over the Imperial Hotel. The presence of the various flags demonstrated some of the diversity that existed within Irish republicanism during the Rising.[6]

Flags are spatial signifiers commonly utilized to demarcate a territorial claim to a place (Nolan et al. 2014), as discussed in Chapter 5. Displaying Irish republican flags during the Rising was a significant symbolic act. In the case of the Irish tricolor flag, it became the national flag of Ireland primary due to its inclusion during that week. Its colors are highly representational: green, commonly associated with Catholics and specifically with Irish republicanism; and orange, often the symbol of Protestantism in Ireland; are bridged by white, the hope for peace between or inclusion of the two sides (see Figure 2.1).

4 The British incorrectly labeled all participants as "Sinn Féiners," after a republican political party that was not involved in the insurrection and would not enjoy widespread public support until after the execution of many of the Rising's leaders (Townshend 2005).

5 The Chief Secretary of Ireland, Augustine Birrell, left Ireland for Easter without arranging any military defense for Dublin Castle. Even after receiving word that Irish rebels stole dynamite on Easter Sunday, British forces planned to respond the following day.

6 For example, while Patrick Pearse (IRB) promoted a violent form of nationalism as a necessary avenue for Irish freedom, James Connolly (ICA), as an Irish socialist challenging oppression by a foreign state, accepted the inevitability of violence in the rebellion but believed it problematic and flawed.

FIGURE 2.1 Irish “tricolor” flag.

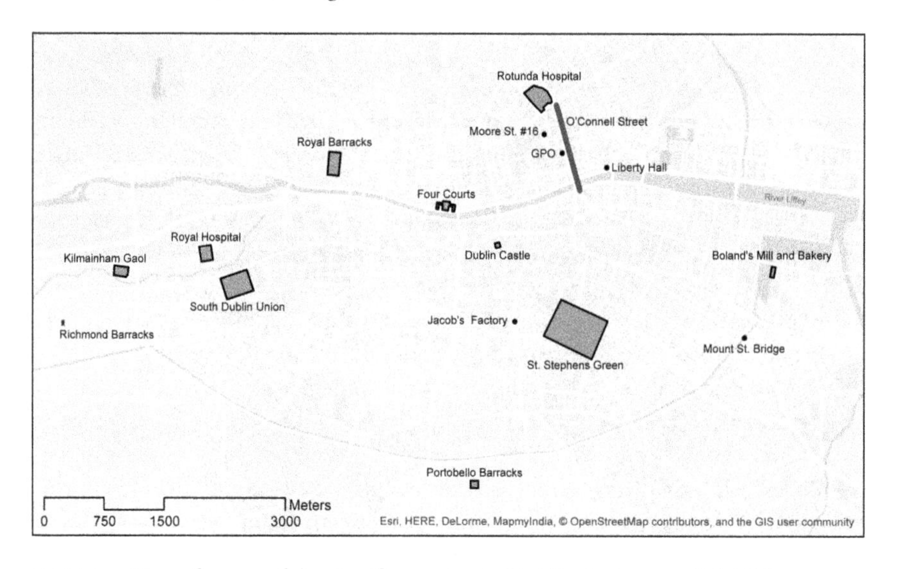

MAP 2.1 Map of many of the significant sites in Dublin during the 1916 Rising

As a result of serving as the platform for the first public reading of the Proclamation, the flag, and a garrison throughout the week, the GPO became a significant place for many within the Republic. However, it is important to note that the perception of the GPO as a nationally sacred space was not instantaneous. Initially, many Dubliners hostilely ridiculed the small group of radicals at the GPO throughout Easter week, as much of the public did not yet share the same geopolitical desires as republicans (McGarry 2010).

In addition to the GPO, the rebels controlled other sites around the city including South Dublin Union, Boland's Mills, Jacob's Factory, Four Courts, Liberty Hall (where they printed propaganda pieces), and St. Stevens Green (see Map 2.1). The rebels took many prominent buildings, not only as tactical platforms from which to publicly promote republican political slogans but also as places that provided high vantage points for their snipers. One exception included St. Stevens Green, an urban park in which republicans constructed trenches and barricades to defend themselves from British attacks launched from Shelborne Hotel and Trinity College (Yeates 2012).

The rebels' goal was to control the city via these key sites long enough to sever government communication throughout the island. They also planned to utilize these locations to disseminate their own propagandized version of the unfolding events. For example, during the Rising, Pearse declared:

> Republican forces everywhere are fighting with splendid gallantry. The populace of Dublin is plainly with the Republic, and the officers and men are everywhere cheered as they march through the streets.
>
> *(Pearse 1916, cited in McGarry 2010)*

Eyewitness accounts, however, suggest that Pearse's report of public support was an exaggeration. Prior to the British response to the Rising, much of the Irish public did not support violent republicanism, which represented only a small portion of political sentiment and Irish nationalism at the time.

At the beginning of the rebellion, violence was highly concentrated in certain parts of the city. As a result, most Dubliners were unaware or confused by the disturbances on the streets and continued with their daily routines throughout the city (e.g., eyewitness accounts, Ceannt 1941). Early in the week, some who were oblivious to the rebellion greeted neighbors in the streets to exchange news about German attacks on the Western Front or discuss the outcome of the Fairycastle horserace. For example, a British army captain, who had just returned from the front, discovered rebels entrenching themselves in St. Stephen's Green while he was walking along Grafton Street. When he ran to his barracks to report what he witnessed, he found his squad was unaware of the unfolding events (eyewitness account, Gerrand 1950).

Once news of the rebellion reached British authorities, they declared martial law in Dublin on Tuesday and cut the city's telephone and telegraph services (except for military use). This severed the public's access to information, thereby increasing confusion. The *Irish Times'* initial news report on the Rising simply stated: "Yesterday morning an insurrectionary rising took place in the city of Dublin" (cited in Hegarty and O'Toole 2006). Thus, with little information and communication lines cut, rumors spread by word of mouth, including some ludicrous suppositions that the rebels captured Cork city or that German forces landed in Ireland.

Destruction and death

To many, the city had been a comparatively quiet space during WWI, so the unfolding events during this week confused, stunned, and enraged many of its inhabitants (Yeates 2012). As news and the realization of the existence of an Irish armed insurrection spread throughout the city, panic began to set in among Dubliners, and chaos quickly ensued (e.g., McGarry 2010). The subsequent eruption of fires and widespread looting on Sackville Street shocked witnesses, while similar disturbances disrupted the city's public transportation and delivery services (Ó Catháin 2008). Dubliners faced widespread food shortages as stores and banks closed and access to grocery stores was limited to a few hours each day.

The British Army, with four British battalions deployed throughout the city, would be responsible for the largest number of civilian casualties, which included children (Ó Catháin 2008). This was primarily due to its willingness to aggressively utilize heavy machine guns and artillery in Dublin. As British reinforcements arrived in the city throughout the week, their shelling and artillery fire intensified the fighting and urban devastation. Public onlookers watched in horror as buildings on the east side of Sackville Street burned and windows melted. The British gunboat, Helga, traveled up the River Liffey (which bisects the city) shelling various buildings, including the aforementioned Liberty Hall.

Other parts of the city witnessed house-to-house style fighting. The non-traditional, sniper-style fighting surprised many of the British forces as they struggled to identify their opponents. As one British soldier who fought against Irish rebels during the Rising explained:

> At the Front, you knew which direction from which you may expect a bullet . . . Here, the enemy is all around you. He lurks in the dark passages and chimney stacks, and when at last you think you have hunted him down, you find yourself in possession of a peaceful citizen who gives some plausible reason for his presence.
>
> *(recorded by Norway 1916)*

The urban devastation varied geographically. For example, Boland Mill saw little action, while the battle for the Mount Street Bridge, only a few hundred yards away, would produce the highest casualties. This was primarily a result of the British troops' employment of traditional trench warfare tactics (see https://mountstreet1916.ie for 3D recreation of the battle). Formally trained to wait in small groups of six to eight men until a whistle signaling a "full frontal charge," the forces who tried to take this bridge faced the rebels' amateur sniper-style attacks from rooftops or apartment windows. An eyewitness described the resulting carnage as "a great pile of dead and dying on the Mount Street Bridge" (Walsh 1949, 18).

When British shelling struck the GPO, it erupted in flame. With their water supply severed, those within the building retreated to nearby rebel-occupied structures. Ultimately, low on munitions and surrounded by troops, the rebels surrendered (Crowley et al. 2017).

Geopolitical implications and legacies of the Rising

Official reports suggest the Rising resulted in 452 deaths (more than half of which were civilian casualties) and injured more than 2,600 people, primarily in Dublin (Townshend 2013). Before the Rising, many in Ireland did not share republicans' desire for national independence, especially as many Irish were overseas fighting with the British against the Central Powers. Initially, most were outraged by the violent actions of the "misguided militant rebels." Dubliners, in particular, were furious about the civilian casualties, destruction and looting, severed gas supplies, disrupted communication lines, and lost business capital. Many also considered the rebels' cooperation with Germany particularly treasonous so recently after the sinking of the Lusitania, which included several Irish among the ship's drowned passengers (McCarthy 2012).

Despite the Irish public's disapproval of the rebels immediately following the rebellion, perceptions of the Rising were soon supplanted by a quasi-religious form of nationalism. Reports of British forces' mistreatment and/or killing of Irish civilians during the Rising shocked the public. Then, instead of treating the rebels as prisoners of war, the state resolved to execute many of the leaders after hasty and secret court-martials. While the police appeared to be indiscriminately arresting thousands of citizens (3,430 men and 79 women) based on contradictory evidence or none at all, some prisoners were executed without trial (Barton 2010).

Patrick Pearse, Thomas MacDonagh, and Thomas Clarke were the first to be executed by firing squad in the Stonebreakers Yard of Kilmainham Gaol (jail). Later, authorities tied the severely injured James Connolly to a chair in the yard before his execution. As news of these proceedings spread, John Redmond, several London-based newspapers, and the British Prime Minister raised concern about the number of executions, fearing it might ignite a retaliatory response among the Irish public. However, the British officer in charge of the trials, Sir John Maxwell, continued the executions. Ultimately, 16 of the original 90 sentenced to death were killed between May 3 and 12, 1916 (Maxwell 1916).

These "excessive British actions" horrified the public (Bonar 1916). The extreme response may reflect the state's indignation over the insurrection and the Irish rebels' willingness to collaborate with the Germans during WWI. However, the state's response to the skirmish, which had little impact beyond Dublin city limits, was perceived by most Irish as vindictive and brutal – a sentiment that would have profound implications for Irish and British geopolitical relations (e.g., McCarthy 2012). Stories of rebels praying to God before their execution, and Pearse's powerful pronouncement that the Rising had been a "holy war" for the Irish nation, fortified the imagery of the rebels as patriotic, national martyrs for many Irish. Soon their struggle gained a cult-like romanticization within the national mythology of Ireland (e.g., Willis 2017).

While a large number of the detained were eventually released, others were deported to prison camps (e.g., in Wales). This further disillusioned the Irish public with British rule in Ireland. As time passed, the Rising inspired many who had previously supported Home Rule to demand independence from Britain. The legacy of the Rising, paired with the state's misguided efforts to impose Irish conscription for WWI, fostered an Irish political collective memory that served as a foundation for the nascent Irish nation. In less than 2 years, republicanism, which was previously considered too radical for many, became the most significant political ideology across "Catholic Ireland."

Commemoration

For many in Ireland, significant sites during the Rising in Dublin have become sacred national space (Barton 2010). For example, the Stonebreakers Yard of Kilmainham Gaol, where many were executed, is now widely regarded as a sacred space due to these individuals' sacrifice for the Irish nation (Patsy McGarry 2016). This and other key sites related to the Rising continue to be politically memorialized through ritual, commemoration, and school tours, as well as in online information guides (e.g., see https://dublinrising.withgoogle.com/welcome/).

The Rising, interpreted as a transformative event, is also interwoven in Irish national lore and widely commemorated in writing and song. For example, Irish poet William Butler Yeats' famous poem, "Easter 1916," memorialized the event. In his poem, he connected the rebels and the color green to republicanism in Irish national sentiment. However, like many in Ireland, he grappled with the guilt of initially mocking the Rising and struggled with misgivings to honor the rebels and the rebellion. He also feared his poem would oversimplify the significance of the event (Kelleher 2017). Nevertheless, Yeats admitted that their self-sacrifice transformed his understanding of the significance of the insurrection in regard to the future of Irish nationalism. As he shared his reluctant admiration and gratitude for the rebels' sacrifice, he explained how their actions altered the geopolitical landscape of Ireland through the birth of a new and embattled Irish nation. As he stated in the last stanza of the poem:

> I write it out in a verse –
> MacDonagh and MacBride
> And Connolly and Pearse
> Now and in time to be,
> Wherever green is worn,
> Are changed, changed utterly:
> A terrible beauty is born.
> (Yeats 1920)

Many of the participants and events that transpired during the Rising are also memorialized in songs (e.g., Off to Dublin in the Green/The Merry Ploughboy, Foggy Dew), theater performances (e.g., the Plough and the Stars), and films/television (e.g., Michael Collins, Rebellion, Ryan's Daughter). Indeed, the perceived martyrdom of the rebel leaders and abhorrence of Britain (especially England) appears in the Ballad of James Connolly:

> God's curse on you, England, you cruel-hearted monster
> Your deeds they would shame all the devils in hell.
> Tere are no flowers blooming but the shamrock is growing
> On the grave of James Connolly, the Irish Rebel!
> Many years have rolled by since that Irish rebellion,
> When the guns of Britannia they loudly did speak.

Te bold IRA they stood shoulder to shoulder,
And the blood from their bodies flowed down Sackville Street.
Te Four Courts of Dublin the English bombarded,
Te spirit of Freedom they tried hard to quell.
For above all the din rose the cry "No Surrender,"
'Twas the voice of James Connolly, the Irish Rebel.
(song lyrics cited in Millar 2020, 74)

Despite the recognition of many of the Rising's participants, research suggests that the role of women is often overlooked or omitted from historical accounts and national commemorations of such events (e.g., McDowell 2008; Valiulis 2011). For example, female members of the ICA, such as Countess Constance Markievicz[7] and her protégé Margaret Skinnider, participated in the Rising (Devine 2013). These women drew inspiration from and were radicalized by the international suffrage movement. Their fervent Irish nationalist ideologies and campaigning efforts included demands for greater inclusion of women in higher education, the professional world, and property/marriage rights in Ireland (e.g., Matthews 2012; Ward 1995). However, despite their participation in this armed struggle and the Irish Proclamation's promise that "Ireland's national right to freedom and sovereignty" includes "the suffrage of all her men and women," recognition of women's contributions in the Irish national struggle remains contentious (e.g., Weihman 2004; Dowler 1998).

Those more critical

While the Rising is a central catalyst in Irish national history, the aforementioned pronouncements of gratitude and recognition were not widely shared by unionists/loyalists in Ireland. The significance of the Rising can also be a more complicated geopolitical event for Irish republicans in Northern Ireland, as the region remains part of the UK today. It is, therefore, important to understand how the geopolitical context in which the event is interpreted strongly influences its territorialized meaning (e.g., Grayson and McGarry 2016). In addition, some Irish in the Republic have challenged the political romanticization of the Rising. Seán O'Casey, a famous Irish playwright, mocked the national "mythology" of the Rising in his controversial satirical work, The Plough and the Stars.

Indeed, how the Rising is remembered and memorialized transforms over time, and is often influenced by current motivations and interpretations (e.g., Willis

7 Countess Markievicz co-founded Na Fianna Éireann (Irish National Boy Scouts), a pseudo-military youth organization that trained and inspired many Irish revolutionaries in 1909 (White 2017).

2017). For example, contemporary Irish author, Anne Enright (2016) is more critical of the Rising as a foundational national myth. As she argues:

> The men of 1916 had no chance of success. They went out to die, a few hundred of them – while thousands were being slaughtered each day in the trenches of Belgium and France. Big wars are terrible and killing civilians is just cowardly but pitching a few hundred men against the might of the British army is a revolution. It was a transcendental moment of sacrifice and of suffering . . . All nations have founding myths . . . the truth is that local history has given way, in my lifetime, to global economics, and we have no good stories for this: no parades, no revolutions. The stories we tell ourselves about the past are not about politics . . . They contain a deeper madness. The stories we tell are about killing and being killed, and why that was all a terrific thing to do.

Significance of growing republican sentiment

The imprisonment and/or execution of rebels and the British state's threat to enforce Irish conscription in WWI in 1918 drastically altered Irish political sentiment. In addition to Britain's response to the Rising, the state's actions laid the political foundations for the Irish War of Independence/Anglo-Irish War. Indeed, the passage of the 1918 Conscription Bill, which threatened to impose an Irish draft for WWI, was a significant factor fueling the public's growing opposition to British governance in Ireland. Irish Volunteers (its successors included the Irish Republican Army that formed in 1919) and Sinn Féin (the political wing of the IRA) exploited anti-conscription fears to bolster their propaganda campaigns across Ireland. When the Catholic Church in Ireland began to support Sinn Féin due to their anti-conscription pledge, the prospect of fighting under the British flag became increasingly disfavored among many Irish. Republicans also touted the "racist attitude of many of British leaders who looked upon the Irish as an 'inferior race'" (Hopkinson 2004, 7) to garner support for their campaign for Irish sovereignty.

British prison camps became breeding grounds for republicanism. In them, prisoners exchanged revolutionary ideas and cemented their desire to again take up arms against the British after their release. Republicans, especially those who had participated in the Rising, organized commemorations for the rebellion on its anniversary (McCarthy 2012). Pride in the Irish language, culture, and sports grew (e.g., Irish hurling and Gaelic football).[8] Eventually, a new, large-scale campaign for an independent Ireland developed – increasingly associated with the previously waning republican party, Sinn Féin. The public also began to emotionally

8 Several political geographers highlight interlinkages between sport and nationalism (e.g., Bleakney and Darby 2018; Koch 2017).

distance their support for the Irish fighting on the Western Front, while British military leaders ostracized Irish troops fighting in the trenches (Morrissey 2005). (Reluctance to honor or commemorate Irish involvement in WWI is discussed in Chapter 8.)

As part of a concerted effort to challenge British geopolitical control of Ireland, the IRA began to intensively recruit and establish local military brigades. They adapted elements from the British army model units that they applied to a parish-by-parish organizational system (i.e., company, battalion, and brigade). They also adopted British military terminology (e.g., dugout, flying columns) as they expanded their presence into the provinces (Borgonovo 2013). By June 1917, the last remaining prisoners from the Rising were released, including Michael Collins and Richard Mulcahy, who would become leaders of the new IRA.

The geographic concentration of republicanism in Ireland at this time was uneven and centralized around local charismatic republican leaders. For example, support for Sinn Féin was strongest in the southwest of Ireland and the northeastern part of Connacht, while weakest in eastern Ulster. Despite the unevenness, growing support for republicanism in Ireland was evident. For example, Sinn Féin's unprecedented and overwhelming victory in the Irish general election of 1918 demonstrates this shift as it defeated the more moderate nationalist IPP, which had previously dominated the Irish political landscape (Townshend 2013).

To protest British rule in Ireland, Sinn Féiners elected for parliament in 1918 refused to take their parliamentary seats in London. Instead, they formed their own separatist parliament, the Dáil Éireann, in 1919. Cathal Brugha served as its first president. The Dáil exemplified a republican counter-state that attempted to undermine British control of the island. Over time, it increasingly focused its efforts on territorial control in 26 of Ireland's 32 counties, commonly referred to as the "south," even though three of the so-called "southern counties" are located in northernwestern Ulster (Townshend 2013).

Two influential republican figures: Collins and de Valera

While there are many republican figures in Irish national lore, few are more prominent than Michael Collins and Éamon de Valera. Collins, who participated in the Rising, was imprisoned and subsequently released. He became Minister of Finance for the Irish Dáil in 1919. That year, he also formed his elite "Squad" that targeted British intelligence agents and detective constables (the "G-men") to cripple British communications in Ireland. He was a charismatic leader, strategist, and public speaker who successfully convinced or coerced many Irish to assist the IRA's fight against the police and British forces. Later, he participated in treaty negotiations with the British state (1921) and eventually served as Chairman of the Irish Free State's Provisional Government in 1922.

During the Irish War of Independence/Anglo-Irish War, he served as the Director of Intelligence for the IRA, a position from which he became well renowned for his guerrilla tactics. In this role, he provided the IRA with arms, financial support, and military strategies. He established an effective network of informers consisting of counteragents, spies, postal workers, railroad employees, hotel porters, and house cleaners who infiltrated the upper echelons of the police, the British government, and its intelligence forces. At the same time, Sinn Féin began to highlight the state's coercive response to the Rising and effectively swayed many to assist the IRA. For example, Ned Broy provided Collins with intelligence that he gathered from the Dublin Metropolitan Police every 24 hours (Mulcahy Papers P7/D/70 1964). While Collins' network initially extended throughout Dublin, he gradually expanded it into the surrounding provinces. He utilized the information that his network gathered to help IRA members with attack strategies and targeted assassinations.

American-born Irish republican Éamon de Valera was a leader during the Rising and was imprisoned for his involvement. After his release, he traveled throughout the United States in an attempt to find financial support and garner recognition for the declared Irish Republic. He also sought international validation for the IRA's armed struggle against British "occupation of Ireland."

de Valera further radicalized Sinn Féin as its president, after its founder, Arthur Griffith, stepped down from leading the party. de Valera eventually left Sinn Féin to form the republican Fianna Fáil party, which would become a powerful driving force during the Irish Free States' development after the Anglo-Irish treaty in 1921 (discussed in Chapter 3). He also served as the President of the Irish Dáil in its second meeting in 1921. He would later serve as the second Taoiseach (Prime Minister), third President of the Republic of Ireland, and strongly influence nationalist narratives and political myths in this nascent Irish state.

Outbreak of the Irish War of Independence/ Anglo-Irish War

Tensions in Ireland grew until the outbreak of the Irish War of Independence/ Anglo-Irish War on January 21, 1919, when the Dáil declared independence from Britain. The politics of nomenclature in Ireland, particularly regarding names of political conflicts, treaties, and geographic locations, often reflect particular ethnonational or political backgrounds. The various names for this war reflect some of the different political entities' competing geopolitical histories. For example, the term Irish War of Independence is more commonly used within the Republic and by those who identify as Irish in Northern Ireland. Republicans sometimes call it the Black and Tan War[9] to highlight the British auxiliaries' atrocities. Those who identify as British commonly refer to it as the Anglo-Irish War. Indeed, this war would be one of competing nationalized narratives, geographic imaginings, and territorial control. The Irish War of Independence/Anglo-Irish War hereafter will be denoted as IWI/AIW.

9 "Black and Tans" is a moniker derived from the force's distinctive uniforms.

During the IWI/AIW, the British parliament passed the Government of Ireland Act of 1920. The act subdivided the island's administration through the creation of two geopolitical areas with devolved parliaments: one in Belfast (representing six of the nine counties in Ulster to ensure that Protestants would possess a political majority) and the other in Dublin (representing the remaining counties on the island). The state passed the legislation without any Irish nationalists/republicans or Ulster unionists[10] voting in its favor. The parliamentary division of the island would be the foundation for Ireland's subsequent official political partition that created Northern Ireland and the autonomous Irish Free State under the Anglo-Irish Treaty of 1921 (Yeates 2012).

War tactics and multiscalar power struggles

On the first day of the war in 1919, the Army of the Irish Republic, the IRA, organized an ambush in Soloheadbeg, County Tipperary, killing two government constables from the Royal Irish Constabulary (RIC). In addition to maintaining law and order, the RIC provided Dublin Castle with intelligence from across the island. The ambush marked the beginning of the IRA's military guerrilla campaign against the RIC, British troops (including the Royal Air Force and Royal Navy), and later, British Auxiliaries forces (i.e., "Black and Tans," Auxiliaries, and the Ulster Special Constabulary). They concentrated their early efforts in Dublin and the southwestern counties (e.g., Cork, Tipperary, Kerry, and Limerick).

The IRA's ambushes represent a geographic and strategic combat style distinct from the formally trained British forces. The non-conventional nature of Irish guerrilla tactics, designed to prevent an opposing army from employing their entire military strength, perfectly suited a smaller, inferiorly armed, and mostly part-time membered IRA. This was particularly strategic as the IRA waged war in its homeland against a far larger, better equipped, and primarily foreign military (Kautt 2014). The British state continued to be reluctant to respond militarily shortly after signing the WWI Paris Peace Conference. Instead, it relied primarily on the RIC to restore order in Ireland (e.g., Townshend 2013.)

The IRA's initial strategy was to target members of the RIC and their barracks. Small IRA battalions raided local RIC barracks to steal arms, ammunitions, and money. These attacks also served as IRA propaganda efforts to demonstrate its success in fighting British rule in Ireland. Research suggests that republican propaganda efforts during the war proved influential (McKenna 2011). Indeed, one of the Dáil's main strategies focused on promoting narratives designed to garner support for their efforts in Ireland and abroad. In 1919, Dáil member and Sinn Féin president Éamon de Valera embarked on a propaganda tour throughout the US.

10 Sinn Féin already abstained from participating in Westminster's House of Commons and unionists did not want to exclude other unionists who would be "left behind" in Catholic Monaghan, Donegal, and Cavan.

During his trip, he condemned the British government's atrocities in the hope of bolstering financial support and publicizing the republican struggle against British rule. In Ireland, republicans proudly heralded the success of IRA ambushes and exploited news of RIC police atrocities. Sinn Féin hung banners and distributed propaganda posters throughout Dublin and other areas with strong republican sentiment, bypassing British censorship of republican news reports and communication. Eventually, the propaganda war, rising Irish patriotic sentiment, and British forces' poor treatment of Irish civilians fomented widespread support for republicanism.

The nature of public support for the IRA intensified as the British state's coercion sharpened the lines of division and strengthened nationalist ideologies across Ireland. While most of the physical combat was confined to IRA and government forces, some Irish civilians contributed to the war effort by assisting republicans in their local community. Other members of the public chose to support the war effort against the British through civil disobedience. For example, Irish railroad employees refused to transport British government troops or supplies throughout Ireland during the war (Hopkinson 2004).

As RIC members became increasingly unpopular in Ireland, their geographic isolation throughout the island made them vulnerable, and the British state eventually ordered the evacuation of many RIC barracks. The RIC withdrawal created a power vacuum in certain locations that republican forces filled, and in many instances set up their own local legislative courts. While the General Headquarters in Dublin officially commanded the entire IRA, individual IRA units often functioned relatively independently at the local level (Borgonovo 2013). The autonomy and geographic distribution of IRA units are significant. They comprised many small, rather self-directed units which frequently launched surprise attacks on enemy forces before "blending in" to disappear into the local landscape or civilian population.

Once the RIC evacuated many of its barracks, the IRA began to ambush British troops and the RIC as they traveled along rural roads. The IRA also targeted government administration to disrupt British coordinating efforts from their centralized location in Dublin Castle. Because IRA units conducted local raids, they had the advantage of knowledge of the local landscape and support of many residents. Eventually, IRA's General Headquarters began directing local units to organize a nationwide operation targeting courthouses, barracks, and tax offices (Townshend 2013). The growing number of localized IRA attacks demonstrated republicans' ability, if even on a small scale, to usurp British rule and self-govern.

As support for republicanism grew in Ireland, Sinn Féin continued to gain a greater electoral foothold throughout the island. By autumn of 1919, the British Parliament in Westminster realized it had underestimated Irish republicanism and the IRA. It elected to supplement the RIC with troops, thereby launching a campaign that gradually increased coercion in Ireland. In response, the IRA stepped up its attacks against the British army and RIC. This resulted in the Crown's subsequent deployment of auxiliary paramilitary groups, including the disreputable

"Black and Tans," who quickly gained an infamous reputation for their indiscriminate retaliation against Irish civilians (Augusteijn 2017).

As British military presence in Ireland increased, the IRA's attack strategy evolved into the formation of "flying columns." These small guerrilla units comprised roughly 100 plain-clothes soldiers who employed various weaponry for rapid mobility and surprise attacks. They also maintained the ability to quickly disappear into the local environment or a public crowd. In contrast, British forces expected "traditional" warfare tactics and openly complained that the IRA did not follow the conventions of war. The flying columns' hit-and-run strategy helped counterbalance some of the tremendous odds that the Irish combatants faced, far outnumbered by British forces. Additionally, the IRA's knowledge of local terrain, support from inhabitants, and access to local military intelligence facilitated their ability to avoid any sizable efforts by British troops. Over time, the sizes of many of the flying columns increased enough to challenge and, at times, prevail against contingents of the Crown's forces (McKenna 2011).

Numerous women bravely contributed to armed conflicts in Ireland. This included Áine Ceannt, Constance Markievicz, Margaret Skinnider, Winifred Carney, Kathleen Clarke, and Lily O'Brennan. However, in contrast to the Easter Rising, the formal role of women's involvement in republican efforts during the IWI/AIW was minimized by patriarchal perceptions. In objection, Hanna Sheehy Skeffington, an executive of Sinn Féin, famously argued:

> An impression exists in some districts that membership of Cumainn [IRA] is confined to men. This is a mistake and every effort should be made to secure [sic] that women shall not only be on the roll of members, but take an active share in the work of Cumainn and the Sinn Féin movement generally.
>
> *(Skeffington, National Library of Ireland 1922, 47)*

During the IWI/AIW, women drove cars to transport combatants and weapons. They also played a key role in republicans' espionage campaign. As the flying columns became more mobile, Cumann na mBan members commonly resupplied the combatants with provisions and organized their temporary housing in various locations.

Female revolutionaries also served as dispatch carriers for IRA communication networks and interrupted RIC and British army communications (Coleman 2003). For example, Lily Merrin famously served as a spy in British-controlled Dublin Castle. Kathleen MacKenna wrote republican propaganda while conducting espionage. The few women jailed for their contributions to the republican cause were often highlighted in propaganda efforts to fuel anti-British sentiment throughout Ireland (Valiulis 1995). Despite their contributions, Constance Markievicz was the only female elected in Ireland to the British Parliament in 1918 and the Irish Dáil in 1919. Only five additional women joined her in the ranks of the second Dáil in 1921.

Geographic concentration of violence in the IWI/AIW

The IWI/AIW was highly geographical in nature. For example, when the British state declared martial law in 1920, it geographically limited its implementation. Also, the uneven spatial distribution of war efforts had several contributing factors, including a noncontiguous patchwork of localized national sentiment (see Kautt 2014). Much of the armed combat was concentrated in Dublin, the province of Munster (focused in County Cork), Western Ireland, and Belfast (O'Halpin 2019). The IRA wanted a significant presence in Dublin to counter the British forces in the city and a platform from which republicans could internationally broadcast their propagandized reports. The city provided republicans access to arms – many of which were stolen and utilized there. There was also a high confluence of republicans in Dublin, including many recently released from prison. The concentration of these individuals and stolen weaponry made their city the first location in which the IRA mobilized for war. Dublin would also serve as the location where the IRA attempted to establish a chain of command.

However, fear of the British intercepting their communications and strong local sentiments made it difficult to gain control in the provincial areas that functioned autonomously. Since these units were highly localized in nature, republican leadership was unsuccessful in their continued efforts to allocate the IRA more evenly throughout the island. Instead, IRA units' formation and operation were commonly administered locally by individuals who wanted to defend their homeland. This was most evident in the province of Munster, which had a long history of Irish national agitation that preceded the formation of the United Irish League in the late 1800s. In addition, several local charismatic republican leaders in Munster inspired and organized several small IRA battalions throughout the province during the IWI/AIW. Despite their sporadic and disjointed occurrences, IRA's guerrilla tactics proved effective.

Other factors influencing the geographically uneven distribution of violence and/or participants included the presence of middling prosperity. This proved to be a significant marker as IRA battalions commonly consisted of young, part-time, middle-class volunteers. Thus, if an area was too impoverished or remote (e.g., Connemara), there was a high likelihood of outmigration and limited involvement in the war.

IRA support and guerrilla warfare were also prevalent in locations where the police and state forces exerted weak governance in the surrounding area. Arguably, the power vacuum worked to promote and instill a rebellious nature in many rural areas. For example, concentrated IRA activity often coincided with areas of agrarian resistance during the late nineteenth and early twentieth centuries and support for Gaelic League, a social and cultural organization that celebrates Irish language, culture, and nationalism (e.g., Smyth 2017).

Sectarianism in Belfast

In contrast to much of Ireland, fighting in Belfast was highly sectarian in nature. With exceptions, most Catholics in Belfast desired an independent Ireland, while most Protestants aligned with unionists/loyalists' demands to remain under the jurisdiction of the British state. Loyalist paramilitaries became increasingly active in Belfast at this time; a Special Constabulary, consisting primarily of Protestants, was established to maintain local order throughout the war. Both forces regularly targeted Catholics and were responsible for nearly 500 deaths in Belfast – most of the victims identified as Catholic. The widespread presence of IRA activities (i.e., Dublin and the south and west of Ireland) fostered a siege mentally among many unionists that eventually contributed to the outbreak of the violent Troubles in Northern Ireland in the late 1960s (Hayes and McAllister 2013).

Another significant geographic difference, when comparing concentration of violence, regarded the Catholic community and the IRA in Belfast which were increasingly enclaved in segregated areas of the city. This led to the IRA battalions in Belfast becoming more defensive, which compounded a sense of being under attack – far more than their counterparts in the south and west of Ireland. In addition, the IRA in Dublin struggled to communicate and coordinate with Belfast's IRA volunteers, thereby further reinforcing Belfast republicans' entrenched and segregated situation. When the border between Northern Ireland and the Irish Free state was established in 1922, the concentration of sectarian violence began to also geographically gravitate to border areas and its surrounding counties, as discussed in Chapter 3 (Augusteijn 2017).

Arrival of the "Black and Tans"

While the IWI/AIW began as a low-level, geographically concentrated guerrilla war across the island, by 1920, the British Parliament moved to enlarge their forces in Ireland. This decision was spurred by a failed assassination attempt on the Lord Lieutenant of Ireland and the state's explicit acknowledgment that it possessed little control over the west and south of Ireland. 1920 would become a key turning point in the war as the death rate rose steeply that year. In January 1920, the state formed and unleashed the force later known as the "Black and Tans." Comprised primarily of ex-soldiers who were ill-informed about Ireland and inadequately monitored, there was little done to prevent or hold them accountable for acts of indiscriminate violence against civilians. Their violence was most concentrated in the south of Ireland, which witnessed the greatest amount of fighting.

As the IRA aggressively responded to this increased pressure, the British state formed the heavily armed "Auxiliaries" in July 1920 and strategically deployed these forces to the 10 Irish counties where the IRA was most prominent (i.e., Dublin and the south of Ireland). Like the "Black and Tans," the Auxiliaries were loathed among the Irish public for their reputation for undiscriminating brutality. For example, these forces were permitted to imprison citizens without trial, execute brutal "unofficial reprisals" against individuals who were suspected supporters or members of the IRA, and retributively sack entire towns (e.g., Cork and Limerick). This ferocity compelled many IRA members to abscond from public life and become full-time combatants. They formed self-organized Irish flying columns dispersed throughout the island, further fueling the violence produced by the conflict (Hopkinson 2004).

Bloody Sunday, 1920

The Auxiliary forces' cruelty outraged and alienated the Irish, and some of the British public, making the war increasingly unpopular as the violence continued to escalate in Ireland. One of the most infamous examples, known as "Bloody Sunday" among many Irish, occurred on November 21, 1920. In retaliation for Collins' Squad and Dublin's IRA brigade's assassination of several British intelligence operatives living in Dublin, members of the RIC and the Auxiliary forces entered a Gaelic football match in Croke Park and indiscriminately fired on the crowd. They succeeded in killing 14 and wounding 65 civilians. They specifically targeted the Gaelic football match because the sport is considered a popular element of Irish culture, and therefore would primarily attract an "Irish" crowd (Bairner 2001). As stated in Chapter 1, Croke Park is also the headquarters of the Gaelic Athletic Association (GAA), which has been transformed in the minds of many Irish as national sacred space. In fact, the GAA's handbook states that the rubble from the 1916 Rising comprises part of their sporting grounds and that the Bloody Sunday attack was the "key event, which allowed the GAA to place itself at the heart of a nationalist myth" (Cronin 1998, 90). As the GAA's official historian, Marcus de Burca explains:

> the GAA was justly proud of the recognition by the British, implicit in the selection of the target for the reprisal, of the Association's identity with what one of the shrewdest of contemporary observers called the "underground nation."
>
> *(1991, 118)*

Despite the presence of competing forms of nationalism in Ireland, many believe there is:

> a distinct spatial dimension to the prevailing discursive politics of Irish national identity enshrined in Croke Park that has emerged, as it had at various times in the past, as a space in and through which Irish national identities are reproduced.
>
> *(Fulton and Bairner 2007, 56)*

In this way, this space is imbued with powerful, highly complex, and often competing forms of national significance for many throughout Ireland.

As the war continued, the state increased the number of military units stationed throughout the island, concentrating primarily in Dublin (Kautt 2014). When British military countermeasures intensified against IRA flying columns, they often geographically dispersed their forces throughout the countryside. This resulted in smaller, violent confrontations and an increase in casualties on both sides of the conflict. In some cases, this violence prohibited authorities' ability to govern large sections of the island. Finally, after the loss of over 1,300 lives and no clear victor, both sides agreed to adjudicate a truce. The war ended in July 1921.

Foundations for continued strife

This chapter explored key geopolitical events, space, and the geography of armed conflict in Ireland during the tumultuous and transformative period between the 1916 Rising and the IWI/AIW. The political and geographic ramifications of these events repositioned geopolitical relations and nationalist narratives in both Ireland and Britain. The conclusion of the IWI/AIW did not resolve tensions between Irish nationalists/republicans and the British state regarding the geopolitical future of the island. Instead, the war only ended when the conflict reached a stalemate.

Despite Westminster's firm belief that its forces would eventually defeat the IRA, republicans' guerrilla tactics sustained the smaller Irish forces during the war thus far, effectively leading to an impasse. When representatives met to negotiate, disagreements continued until the state pressured the Irish delegation to accept what would prove to be an extremely controversial treaty. Unfortunately, the "Articles of Agreement for a Treaty Between Great Britain and Ireland," more commonly referred to as the Anglo-Irish Treaty, did little more than lay the foundation for more bloodshed on the island. Instead of instilling peace in Ireland, the treaty sowed the seeds of partitioned national ideologies, discord, and intractable violence. The following chapter focuses on the geographic and political implications of the treaty, with particular attention paid to legacies of division, partition, and adversarial nationalist narratives.

References

Augusteijn, Joost. 2010. *Patrick Pearse: The Making of a Revolutionary*. London: Springer.

———. 2017. "Military Conflicts in the War of Independence." In *Atlas of the Irish Revolution*, edited by John Crowley, Donal Ó Drisceoil, and Mike Murphy. New York, NY: New York University Press.

Bairner, Alan. 2001. *Sport, Nationalism, and Globalization: European and North American Perspectives*. Albany, NY: Suny Press.

Barton, Brian. 2010. *Secret Court Martial Records of the Easter Rising*. Cheltenham: The History Press.

Bleakney, Judith, and Paul Darby. 2018. "The Pride of East Belfast: Glentoran Football Club and the (Re) Production of Ulster Unionist Identities in Northern Ireland." *International Review for the Sociology of Sport* 53 (8): 975–96.

Bonar, Andrew. 1916. *"The Bonar Law Papers." April 1916*. House of Lords Record Office. BL/ 50. London: Parliamentary Archives.

Borgonovo, John. 2013. *The Dynamics of War and Revolution: Cork City, 1916–1918*. Cork: Cork University Press.

Búrca, Marcus de. 1991. *The Story of the GAA to 1990*. Dublin: Wolfhound Press.

Callanan, Frank. 2017. "The Home Rule Crisis." In *Atlas of the Irish Revolution*, edited by John Crowley, Donal Ó Drisceoil, and Mike Murphy. Cork: Cork University Press.

Ceannt, Áine. 1941. "Correspondence with the Bureau of Military History." National Library of Ireland. Ceannt and O'Brennan Papers, 1851–1953.

Coleman, Marie. 2003. *County Longford and the Irish Revolution, 1910–1923*. Co. Kildare: Irish Academic Press.

Cronin, Mike. 1998. "Enshrined in Blood the Naming of Gaelic Athletic Association Grounds and Clubs." *The Sports Historian* 18 (1): 90–104. https://doi.org/10.1080/17460269809444771.

Crowley, John, Donal Ó Drisceoil, Michael Murphy, John Borgonovo, and Nick Hogan. 2017. *Atlas of the Irish Revolution*. New York, NY: New York University Press.

Devine, Francis. 2013. *A Capital Conflict: Dublin City and the Lockout*. Dublin: Four Courts.

Dowler, Lorraine. 1998. "'And They Think I'm Just a Nice Old Lady' Women and War in Belfast, Northern Ireland." *Gender, Place & Culture* 5 (2): 159–76. https://doi.org/10.1080/09663699825269.

Enright, Anne. 2016. "Colm Tóibín, Anne Enright, Roddy Doyle . . . The Easter Rising 100 Years On." *The Guardian*, March 26, 2016. www.theguardian.com/books/2016/mar/26/easter-rising-100-years-on-a-terrible-beauty-is-born.

Ferriter, Diarmaid. 2015. "Diarmaid Ferriter: Why the Rising Matters." *Irish Times*, September 23, 2015. https://www.irishtimes.com/culture/heritage/1916-schools/diarmaid-ferriter-why-the-rising-matters-1.2353812.

Fulton, Gareth, and Alan Bairner. 2007. "Sport, Space and National Identity in Ireland: The GAA, Croke Park and Rule 42." *Space and Polity* 11 (1): 55–74.

Gerrand, E. 1950. *Defence of Beggars Bush Barracks by British Easter Week, 1916*. Statement by Witness. Dublin: Bureau of Military History.

Grayson, Richard S., and Fearghal McGarry. 2016. "Introduction." In *Remembering 1916: The Easter Rising, The Somme, and the Politics of Memory in Ireland*, edited by Richard S. Grayson and Fearghal McGarry, 1–9. Cambridge: Cambridge University Press.

Hayes, Bernadette, and Ian McAllister. 2013. *Conflict to Peace: Politics and Society in Northern Ireland over Half a Century*. Manchester: Manchester University Press.

Hegarty, Shane, and Fintan O'Toole. 2006. *The Irish Times Book of the 1916 Rising*. Dublin: Gill and Macmillan.

Hopkinson, Michael. 2004. *The Irish War of Independence*. Montreal: McGill-Queen's Press-MQUP.

Johnson, Nuala C. 2003. *Ireland, the Great War and the Geography of Remembrance*. Vol. 35. Cambridge: Cambridge University Press.

Kautt, William Henry. 2014. *Ground Truths: British Army Operations in the Irish War of Independence*. Co. Kildare: Irish Academic Press Ltd.

Kelleher, Margaret. 2017. "Literary Revival." In *Atlas of the Irish Revolution*, edited by John Crowley, Donal Ó Drisceoil, and Mike Murphy. New York, NY: New York University Press.

Koch, Natalie. 2017. *Critical Geographies of Sport: Space, Power and Sport in Global Perspective*. 1st ed. Oxfordshire: Taylor & Francis.

Mansergh, Martin. 2017. "Ulster's Solemn League and Covenant, 1912." In *Atlas of the Irish Revolution*, edited by John Crowley, Donal Ó Drisceoil, and Mike Murphy. New York, NY: New York University Press.

Matthews, Ann. 2012. *Dissidents: Irish Republican Women 1923–1941*. Cork: Mercier Press Ltd.

Maxwell, John G., and Herbert Asquith. 1916. "The Irish Rebellion." In *Report on the State of Ireland Since the Rebellion*. Richmond: The National Archives.

McCarthy, Mark. 2012. *Ireland's 1916 Rising: Explorations of History-Making, Commemoration & Heritage in Modern Times*. London: Ashgate.

McDowell, Sara. 2008. "Commemorating Dead 'Men': Gendering the Past and Present in Post-Conflict Northern Ireland." *Gender, Place & Culture* 15 (4): 335–54. https://doi.org/10.1080/09663690802155546.

McGarry, Fearghal. 2010. *The Rising Ireland: Easter 1916*. Oxford: Oxford University Press.

———. 2011. *Rebels: Voices from the Easter Rising*. Dublin: Penguin.

McGarry, Patsy. 2016. "Kilmainham Gaol a 'Sacred Place,' Says Archbishop Martin." *Irish Times*, October 24, 2016. www.irishtimes.com/news/social-affairs/religion-and-beliefs/kilmainham-gaol-a-sacred-place-says-archbishop-martin-1.2841456.

McKenna, Joseph. 2011. *Guerrilla Warfare in the Irish War of Independence, 1919–1921*. London: MacFarland and Co.

Millar, Stephan. 2020. *Sounding Dissent: Rebel Songs, Resistance, and Irish Republicanism*. Ann Arbor: University of Michigan Press.

Morrissey, John. 2005. "A Lost Heritage: The Connaught Rangers and Multivocal Irishness." In *Ireland's Heritages: Critical Perspectives on Memory and Identity*, edited by Mark McCarthy, 78–79. Aldershot: Ashgate.

Mulcahy Papers. 1964. "Mulcahy's Notes on the Second Volume of Piaras Béaslai's Biography of Collins." Dublin, Mulcahy Papers, U.C.D. Archives.

Nolan, Paul, Dominic Bryan, Clare Dwyer, Katy Hayward, Katy Radford, and Peter Shirlow. 2014. *The Flag Dispute: Anatomy of a Protest*. Belfast: Queen's University Belfast.

Norway, Mrs Hamilton. 1916. *The Sinn Fein Rebellion as I Saw It*.

Ó Catháin, Máirtín. 2008. *A Land Beyond the Sea: Irish and Scottish Republicans in Dublin, 1916*. Dublin: Irish Academic Press.

O'Halpin, Eunan. 2001. "The Geopolitics of Republican Diplomacy in the Twentieth Century." In *Paper Presented to the IBIS Conference "From Political Violence to Negotiated Settlement: The Winding Path to Peace in Twentieth Century Ireland,"* University College Dublin, 23 March 2001. Dublin: University College Dublin. Institute for British-Irish Studies.

———. 2019. "PJ Moloney's 1916 Journal and Introduction." *Tipperary Historical Journal* 32: 132–53.

Sheehy-Skeffington, Hanna. 1922. *Minute Book of Cumann Na Teachtaire (League of Women Delegates)*. Sheehy-Skeffington Collection, Dublin: National Library of Ireland.

Skinnider, Margaret. 1917. "Doing My Bit for Ireland." *New York, Century*.

Smyth, William. 2017. "Nineteenth-Century Ireland: Transformed Contexts and Class Structures." In *Atlas of the Irish Revolution*, edited by John Crowley, Donal Ó Drisceoil, and Mike Murphy, 4–55. New York, NY: New York University Press.

Townshend, Charles. 2013. *The Republic: The Fight for Irish Independence, 1918–1923*. City of Westminster, London: Penguin.

Valiulis, Maryann Gialanella. 1995. "Power, Gender, and Identity in the Irish Free State." *Journal of Women's History* 7 (1): 117–36. https://doi.org/10.1353/jowh.2010.0308.

———. 2011. "The Politics of Gender in the Irish Free State, 1922–1937." *Women's History Review* 20 (4): 569–78.

Walsh, Thomas. 1949. *Mount St. Bridge Defence, Easter Week 1916*. Statement by Witness. Dublin: Bureau of Military History.

Ward, Margaret. 1995. *In Their Own Voice: Women and Irish Nationalism*. New York, NY: Atrium.

Weihman, Lisa. 2004. "Doing My Bit for Ireland: Transgressing Gender in the Easter Rising." *Éire-Ireland* 39 (3): 228–49. https://doi.org/10.1353/eir.2004.0025.

White, Gerry. 2017. "'They Have Rights Who Dare Maintain Them': The Irish Volunteers, 1913–15." In *Atlas of the Irish Revolution*, edited by John Crowley, Donal Ó Drisceoil, and Mike Murphy. New York, NY: New York University Press.

Wiel, Jérôme aan. 2017. "Ireland and the Bolshevik Revolution." *History Ireland* 25 (6): 38–42.

Willis, Claire. 2017. "Staging the Rising." In *Atlas of the Irish Revolution*, edited by John Crowley, Donal Ó Drisceoil, and Mike Murphy. New York, NY: New York University Press.

Yeates, Paidraig. 2012. "A City in Wartime: Dublin 1914–1918 (Dublin, 2011); P. Yeates." *A City in Turmoil: Dublin 1919–1921*: 1912–23.

Yeats, William Butler. 1920. *Easter, 1916*, Vol. 25, 69. New York, NY: The Dial.

3

PARTITION AND DIVISION

The Anglo-Irish Treaty, Civil War, and Ireland divided

This chapter investigates geographic and political divisions over territorial control and governance in Ireland after the IWI/AIW. It analyzes controversies within Irish nationalism regarding the Anglo-Irish Treaty and the resultant Irish Civil War. It also traces statecraft strategies and ramifications of the creation of the Irish Free State, Northern Ireland, and the partition of Ireland.

As this chapter demonstrates, the terms of the Anglo-Irish Treaty were quite controversial, particularly for Irish republicans. Despite the Dáil's acceptance of the treaty, Irish–British tensions lingered. Depictions of this discontent are evident in books, films, and music from/depicting the era, including the 1996 film *Michael Collins*. In a brief, poignant exchange, Collins responds to a British officer's derisive comment about his late arrival at the ceremony recognizing the Irish Free State's new leadership:

British officer: You are seven minutes late, Mr. Collins.
Collins: You kept us waiting for 700 years. You can have your seven minutes.

Collins' response invokes the territorialized memory of British colonialism of Ireland and struggles for nationhood. The film also portrays some of the binary socio-spatial constructions of Irish identity and homeland imagined along "anti-British" lines in several examples, including another scene with Collins in which he demands:

Collins: Give us the future, we've had enough of your [Britain] past. Give us back our country to live in, to grow in, to love.

In this quote, Collins evokes landscape and place-making as necessary nation-building projects for the future of Ireland, one that, according to Collins' statement, should be free of its geopolitical ties to Britain.

DOI: 10.4324/9781003141167-3

This film also explores dichotomies forged among republicans over the treaty. Debates regarding sovereignty and partition fueled turmoil that soon erupted into civil war in which Collins and pro-treaty republicans, backed by the British state, fought de Valera and anti-treaty republicans. The war not only cost Collins his life but exposed political fault lines among republicans that persisted for decades. This chapter examines how these divisive forces – geopolitical practices (internal and intra-island divisions) and exclusionary nationalist narratives – shaped Ireland and the socio-political paths of its two distinct partitioned political spaces. The subsequent rifts produced sentiments of difference, alienation, and strife that fomented violence across Ireland.

Internal divisions over Ireland's geopolitical future

As discussed in Chapter 2, Ulster unionists openly challenged Irish Home Rule, fearing it would eventually sever their political link with Britain. However, as a minority in Ireland, they anticipated their inability to block Home Rule and petitioned the state for an autonomous government in which they would be the ruling majority. As a result of their fervor, the state presumed a single political entity would not be sustainable on the island under the weight of Ireland's bipartisan divisions. In 1920, the government capitulated and passed the Government of Ireland Act. This legislation nominally partitioned the island into two political entities administered by the two devolved governments – "north" and "south."

In order to forge a territory with the largest possible Protestant unionist majority that would be economically viable, the state partitioned the province of Ulster between the three primarily "Catholic/nationalist" counties in the northwest (Donegal, Cavan, and Monaghan) from the four primarily "Protestant/unionist" northeastern counties (Londonderry, Antrim, Monaghan, Armagh, and Down). The northern unit also included the two additional counties (Tyrone and Fermanagh) that had a Catholic/nationalist majority. Thus, a six-county "northern" political unit (six of the nine Ulster counties) that preserved a Protestant/unionist majority (2:1 unionist ratio) was nominally partitioned from the "southern" 26 counties (9:1 nationalist ratio) (see Map 3.1).

After partition, representatives of the British state, the Irish northern unit (leader – Protestant unionist, James Craig), and the Irish southern unit (leader – Catholic republican, Éamon de Valera) met in secret to discuss the territorial repositioning of the island's partition. Unable to reach a new arrangement, Irish partition continues to be an omnipresent component of British–Irish geopolitical relations.

In the northern unit of partitioned Ireland, unionist "siege mentality" shaped sectarian laws, underpinned poor treatment of Catholics, and fomented violent episodes in the 1920s. Concerned by the increased sectarian violence, Collins tried to negotiate two political agreements with the northern unit's leader to help protect Catholics in the north. However, neither side observed these agreements, and violence continued to erupt periodically in the subsequent decades. The turmoil illustrates how the reterritorialization of geopolitical space, particularly through

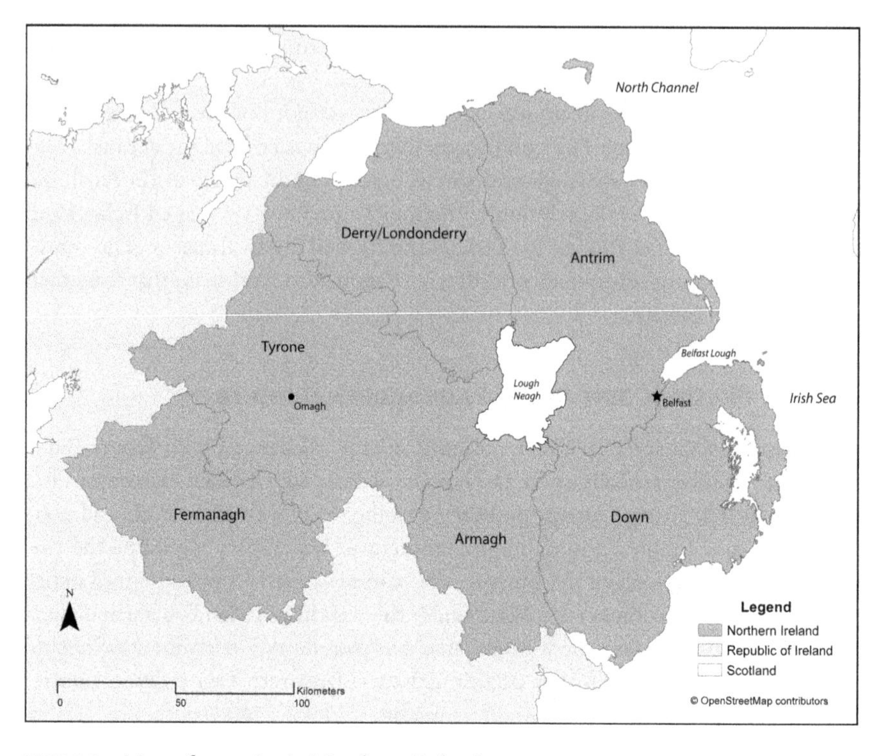

MAP 3.1 Map of counties in Northern Ireland

bordering, exacerbates internal political fractures in society and has increased bloodshed in Ireland. Indeed, the Irish Civil War and Northern Ireland's turbulent era, the Troubles, are only some of the violent ramifications of this partition (e.g., Foster 2015).

Geographic understanding of partition and borders

Partition is a territorial and geographic tactic utilized by external states, colonial powers, global security regimes (such as the United Nations), or imperial forces to divide a territory into two (or more) political entities. The belief in the existence of an inexorable "primordial" difference between contentious groups drives this geopolitical reconfiguration of space (e.g., Rankin 2007). Indeed, the partition of Ireland occurred within the geopolitical context of Britain's imperial governance policy to divide "two nations" within a single territory (similar to the subsequent partitions of India or Palestine). While the intent may have aimed to ameliorate political, cultural, or ethnic turmoil, the tactic addresses the symptoms of eth-nonational tensions rather than offering a solution to the conflict. Additionally, the location of a partition is "hardly ever satisfactory and almost always problematic," and regularly intensifies conflict instead of resolving it (Waterman 1987, 160). For

Ireland, it increased sectarianism and violence; the loss of lives attributed to partition included 544 deaths between 1920 and 1922 and over 3,000 between 1968 and 1998 (O'Leary and McGarry 2016, 21).

Borders thus may best be interpreted as a *process* instead of a static line mapped on the landscape. The complex practices and multiscalar networks that constitute a border area – or "borderscape" – transcend a fixed line (e.g., Brambilla and Jones 2020). Additionally, a border's function, political status, symbolic meaning, and permeability change over time (e.g., Johnson et al. 2011). Scholars of border studies also examine selected and differential "mobilities across" a frontier, dynamics of power among sovereign states, technological regimes, local resistance and counter agency, and cross-border communities (e.g., Komarova and Hayward 2019).

The socio-political processes of border-making or "bordering" emerges through and is sustained by political narratives, representations in the media, and inhabitants' daily socio-spatial interactions within a borderscape. This analytical lens fosters a greater understanding of the spatiality of ethnonational conflicts and politics of borderscapes in Ireland. Additionally, this framework can provide insight into the lasting impact of the partition that followed county lines instead of significant ethnonational boundaries. In effect, the partition became a spatial boundary of dispossession, displacement, and alienation (Dodds 2013). The territorial nature of border enforcement, particularly through categorizations of "belonging" or "foreign," created new minorities located on the "wrong side" of a line. It also tore communities apart through an artificial construction of difference, despite commonalities throughout the border region (e.g., Dempsey 2022).

Disparate perceptions of Irish partition

While the 1920 partition was provisional, its creation set a geographic and political precedent. Once formed, the northern devolved government quickly instituted governance legislation that changed British voting laws and established its own local police force. This political machinery worked to institutionally reinforce partition through the presence of an "established" regional government. Thus, when this northern government was invited to participate in the Anglo-Irish Treaty negotiations in 1921, James Craig argued that the 1920 Act was their "final settlement" (Lee 1990).

In contrast, nationalists/republicans perceived the partition as temporary and negotiable. Two of the Irish Dáil's most imperative aspirations for the treaty included repositioning the political status of six northern counties that comprised the devolved northern territory. The other was the establishment of a path for an independent Irish republic. Other issues, such as defense and control of Irish ports, were of lesser importance to republicans at the time. When British Prime Minister David Lloyd George refused to grant the Dáil a republic – particularly as that would set a precedent for other British colonies to emulate – he asked the northern government if it would agree to be controlled by the Dáil in Dublin. Unsurprisingly, James Craig refused to accept this proposed compromise. This prompted the

British state to threaten creating a Boundary Commission (discussed later in this chapter) to re-draw the border between the two devolved governments should Craig refuse to consider negotiations. While the Border Commission did not ultimately alter the 1920 boundary, Lloyd George's warning to Craig led Irish republican Arthur Griffith, who participated in the 1921 treaty negotiations, to assume that the 1920 partition was temporary. More specifically, he believed the northern unit's territory would be reduced, limiting its economic viability until its eventual reunification with the "southern" unit in Ireland. His mistaken assumption influenced his interpretations of Lloyd George's arbitration tactics and subsequently his own strategy throughout the treaty negotiations (Heintz 2009).

Contentious treaty negotiations

Tensions among republicans regarding the geopolitical future of Ireland after the IWI/AIW were high. Amid mounting debates, Éamon de Valera, the acting president of Sinn Féin, appointed members to a delegation that represented the Dáil in treaty negotiations in London in 1921. Arthur Griffith and Michael Collins led this Irish delegation, while de Valera remained in Dublin. de Valera argued that his continued presence in Ireland was indispensable for preparing the Dáil's republican "hardliners" to accept any treaty that Griffith and Collins negotiated with the state. However, Collins strongly opposed his own appointment to the delegation. He believed his brusque personality and military background leading the IRA and IRB made him unsuitable for diplomatic negotiations. He also argued that because he and Griffith were not provided with specific instructions before they embarked on the negotiations, his delegation was sent to London as scapegoats to do what those in the Dáil, "knew must be done, but had not the moral courage to do themselves" (Collins 1922, 168). This statement highlights the presence of strong internal tensions among republicans that predated the signing of the treaty, which only intensified strife until the outbreak of civil war.

The lengthy treaty negotiations were highly contentious, particularly as Collins and Griffith realized that the state would not grant them a republic or eliminate the partition in Ireland. Instead, it offered dominion status within the British Commonwealth, a concession the delegation loathed to accept. Lloyd George denied them the ability to discuss negotiations with Dáil, further intensifying tension and confusion among republicans on both sides of the English Channel. War-weary Collins expressed his frustrations when he explained:

> I will not agree to anything which threatens to plunge the people of Ireland into a war . . . Still less do I agree to being dictated to by those [the Dáil in Ireland] not embroiled in these negotiations.
>
> *(Collins, undated voice recording during negotiations)*

Once negotiations ended, PM Lloyd George threatened renewed warfare should the Irish delegation reject the offered treaty. Both Griffith and Collins agreed to sign the treaty believing it represented the "best offer" they would receive, with the virtue of avoiding a return to war. However, the delegation feared how the Irish public, the Dáil, and especially the Irish army would respond to the treaty. Collins was discouraged by what the treaty offered, famously stating he had just "signed my death warrant" by approving the treaty. However, he also believed it provided a political path to obtaining a republic, or in his words, "the freedom to achieve it" (Collins 1922).

The Anglo-Irish Treaty (1921): reinforcing partition and divisions

The "Articles of Agreement for a Treaty between Great Britain and Ireland," commonly known as the Anglo-Irish Treaty, offered the 26 "southern" counties self-governing dominion status as the Irish Free State (Saorstát Éireann) in the British Commonwealth (equivalent to Canada or Australia). This marked the end of 120 years of direct British rule for 26 of the 32 counties in Ireland. While this Free State received more political autonomy within the British Commonwealth than if Home Rule had passed in 1914, the Free State was not a republic. As the British king remained the Head of State, the Irish parliament in the Free State was required to take an oath to his allegiance. The state also controlled a number of Irish ports[1] for security purposes and preservation of trade routes (Kissane 2005).

The treaty also reinforced the preexisting partition of Ireland by permitting the remaining six northern counties to elect to leave the newly formed autonomous Free State. The northern unit utilized its ability to leave the Free State within one month of the treaty's ratification. As a newly constituent part of the United Kingdom, "Northern Ireland" included a devolved parliament at Stormont in Belfast.

When the Dáil received the Anglo-Irish Treaty, many republicans, including de Valera, strongly opposed the agreement. In addition to noting the treaty fell short of establishing an Irish republic, detractors refused to take an oath to the king. Indeed, the oath became one of the most divisive debates among republicans. The continued partition of Ireland was also of great concern, but the primary focus remained on the geopolitical status of the Free State. During the three weeks in which the Dáil discussed the treaty, politicians, institutions, and the public began to divide into two camps: those who supported the treaty (pro-treaty) and those who opposed it (anti-treaty). After lengthy and contentious debates, the Dáil narrowly voted to approve the treaty on January 7, 1922, with Collins as the first Chairman of the Provisional Government (hereafter PG).

The geographic, economic, and socio-political ramifications of this treaty are numerous and far-reaching. For example, before EU integration, the border evolved as an economic frontier that inhibited trade and cross-border labor markets within Ireland (Shuttleworth 2007). Additionally, tensions among republicans (i.e.,

1 The British state relinquished control of the Irish Treaty Ports in 1938, allowing the Irish Free State to maintain neutrality during WWII (Moody and Martin 2001).

pro-treaty versus anti-treaty) continued to escalate over the Free State's dominion status and the creation of Northern Ireland. The majority of pro-treaty advocates believed the partition was temporary and that Northern Ireland could not endure over time. Many who supported the treaty were also weary of war. In contrast, the anti-treaty camp refused to accept the Free State's compromised geopolitical status, interpreting it as a betrayal of republicanism. Unreconciled, violence erupted among republicans within 6 months of the treaty's confirmation.

The Irish Civil War (1922–1923)

The Irish Civil War began in Dublin on June 28, 1922, when the PG attacked anti-treaty militants that garrisoned their forces in the city's court building, the Four Courts. Anti-treaty Irish, commonly labeled as "Republicans" or "Irregulars," created an opposition government with Éamon de Valera as its president. However, Republicans in Northern Ireland functioned independently throughout the war. The utilization of the term *Republican* during the Irish Civil War (i.e., the remaining portions of this chapter) specifically refers to anti-treaty forces, rather than the political affiliation described throughout the rest of this book.

As the Chairman of the British-backed PG,[2] Commander-in-Chief of the National Army, and head of the IRB, Collins was pressured by the state to suppress resistance to the treaty and attack these garrisoned Republican forces. The resultant 10-month civil war split nationalists, republicans, Sinn Féin, and the IRA between pro-treaty and the anti-treaty sides (e.g., pro-treaty IRA and the National Army of the Free State, versus anti-treaty IRA). Many pro-treaty combatants joined as a result of their loyalty to Collins or to a local IRB unit. While much of the IRA leadership supported the treaty, estimates suggest that almost 75% of the enlisted IRA opposed the treaty (Borgonovo 2011). The treaty also divided Cumann na mBan (Irish Women's Council) into pro-treaty Cumann na Saoirse and anti-treaty Cumann na mBan.

The public generally supported the superiorly equipped and financed PG forces. The PG's ability to pay its military effectively recruited many enlistees, and the Catholic Church and most media sources spread pro-treaty propaganda. Most Protestants in the Free State were pro-treaty. However, while most affluent supported the treaty (e.g., businesses, large farm owners, commercial press, middle class, and the Catholic Church), smaller agrarian workers and laborers, especially those in rural Western Ireland, commonly opposed the agreement (Foster 2015).

Throughout the war, Protestants in the Free State were targets of sectarian violence underpinned by exclusive socio-political bordering narratives that framed Protestants as "foreign" in the Free State. Despite the British state's promise to

2 The state provided the PG with weapons, munitions, and financial aid as "counter insurgency support."

formulate an evacuation plan for these Protestants, it failed to follow through and left them with little protection throughout the war (Foster 2015). Evidence suggests that sectarian attacks drove a large number of Protestants from the Free State during the civil war (Bielenberg 2013).

Sectarian violence against Protestants in the Free State further exacerbated Northern Ireland's Protestant/unionists sense of vulnerability and "siege mentality" as a targeted minority in Ireland (e.g., McAuley 2016). Their reactive political and cultural measures against Catholics/nationalists subsequently increased in the north. Unionists' targeted methods of oppression of the Catholic/nationalist community contributed to decades of violence in Northern Ireland and the outbreak of the Troubles in the late 1960s (discussed in Chapters 4–6).

Geographic concentration of violence

The violence during the civil war was geographically concentrated in Dublin and the province of Munster (especially in Kerry, Cork, Limerick, and Tipperary counties). Anti-treaty support was highest in the west of Ireland (e.g., Hopkinson 2010). Analogous to the IWI/AIW, IRA units remained highly localized and principally functioned independently. Accordingly, the geography of IRA units' allegiance varied greatly as individual IRA organizations and their members split along pro- and anti-treaty lines.

There is a geographic correlation between areas of heavy combat during the IWI/AIW, such as Cork and Tipperary, and anti-treaty forces/support during the civil war. This may be a result of their bellicose contributions during the IWI/AIW, which provided locals with combat experience and fostered a sense of localized self-sufficiency. Many IRA members in Cork and south Tipperary also believed that they were not sufficiently supported by Dublin during the IWI/AIW and felt excluded from the Anglo-Irish Treaty negotiations. In addition to their remote location and distance from Dublin, these perceptions contributed to a distrust of the pro-treaty PG in Dublin (Hopkinson 2010).

At the onset of the war, anti-treaty Republicans held a numerical advantage (12,900 members) over the PG's forces (6,000 armed members). Archival evidence suggests, with PG forces concentrated in Dublin, Republicans controlled much of the west of Ireland and offered a "home territory" advantage for much of the war (MacMahon 1924). However, not all of the newly enlisted Republicans' were highly skilled or well-trained in combat. Obtaining arms and ammunition throughout the war also proved problematic for Republicans, unlike their British-backed PG opponents. As their supply chains dwindled, some anti-treaty units resorted to raiding local civilians, souring much of their support. Also, not every Republican wanted to return to war. As Cork Republican Florrie Begley explained:

> We had no heart in the Civil War . . . When the attack on the Four Courts was over, the fight should have stopped.
>
> *(cited in Collins 1922 S 9241)*

As the larger and superiorly outfitted PG forces continued to expand their numbers (60,000 by April 23) and defeat or capture their opponents, anti-treaty forces increasingly relied on guerrilla tactics. While these attacks were effective, particularly in mountainous areas and through the destruction of railways in the west and southwest of Ireland, the impact was geographically limited. In response, the PG passed wide-sweeping and severe internment and/or execution policies that resulted in atrocities that further embittered Irish politics.

By 1922, much of the anti-treaty IRA was forced underground. In response, the Cumann na mBan (Irish Women's Council) emerged to organize and lead Republican protests, disseminate anti-treaty propaganda, and care for injured combatants. In contrast to the small number of female prisoners taken during the IWI/AIW, more than 560 women were imprisoned during the civil war (McCoole 2004). Despite their participation in politics and bravery in warfare, the role and opportunities for women increasingly diminished in the Free State as a result of male efforts to marginalize women in public (e.g., Borgonovo 2011; McCoole 2004; McDowell 2008).

Collins' death, a turning point

Collins was killed in an ambush on his convoy in County Cork on August 22, 1922. As the only fatality in a group that traveled to Cork to discuss potential peace talks, his death sparked several conspiracy theories. Archival evidence suggests Collins' death marked a turning point for the Free State government's willingness to implement more ruthless tactics against anti-treaty Republicans (Belton 1922). One of the most infamous new policies of the PG, now led by W.T. Cosgrave, included the rapid execution of war prisoners, many without trial (Ferriter 2015). The PG implemented this tactic to deter Republican attacks, often agreeing to stay executions if the anti-treaty forces agreed to local ceasefires. While this approach proved effective, it did little to improve public opinion of the PG's increasingly violent actions during the war (Hopkinson 2010). The capability of Republican forces continued to erode, until their leader Liam Lynch died in combat. His replacement, the more pragmatic Frank Aiken, eventually agreed to a ceasefire in April 1923 and the Irish Civil War concluded on May 24, 1923.

Placing blame

After the war, exclusionary nationalist narratives were not limited to partisan perceptions that pitted British versus Irish. Historically, many pro-treaty supporters believed de Valera was responsible for the outbreak of civil war, claiming he instigated and spread anti-treaty furor in Ireland. In contrast, anti-treaty factions blame Collins and Griffith for accepting the treaty without consulting other members of the Dáil. However, while de Valera publicized anti-treaty sentiments and was the nominal head of Republicans, his *de facto* control was limited. For example, he was not consulted about the 1922 occupation of Dublin's Four Courts and disliked the intensely autonomous nature of the Republican IRA units throughout the war (Foster 2015).

Ultimately, the Irish Civil War was a struggle for *de facto* geopolitical and fragmented sovereign control of sections of Ireland. Indeed, territorial challenges to sovereign claims are the geopolitical foundations of civil wars. Anti-treaty forces believed theirs was an anti-colonial struggle against the British-backed Irish PG, while the government saw Republicans as an impediment to an eventual republic. After cessation of hostilities, divisions persisted until some anti-treaty Irish left the Free State in the mid-1920s; known as the "Wild Geese," many settled in the US, England, Canada, and Australia (Foster 2015).

An entrenched border

After its creation, the border rapidly became entrenched in Ireland. The outbreak of civil war diverted attention from negotiations and postponed the Boundary Commission. Collins' death was a great loss for advocates of removing the partition (despite the paradox that the treaty reinforced the unwanted border). Instead of economic cooperation between the two political units in Ireland, the Free State focused its trade with Great Britain and forged a customs barrier against Northern Ireland in 1923 (MacDonald 2018).

W.T. Cosgrave, who served as the Free State's President of the Executive Council from 1922 to 1932, insisted on a Boundary Commission to discuss the future of the partition. However, the previously "established" partition helped PM Lloyd George to coerce the Free State government to accept the existence of the Irish border in 1925. When he dissolved the Irish Boundary Commission, the border became entrenched (see Rankin 2007).

The Irish Free State

In this nascent political space, debates over the future of the Irish nation continued. In May 1923, the public elected W.T. Cosgrave's pro-treaty party, Cumann na nGaedheal to office. Its electoral majority in the 1920s facilitated the new government's ability to implement the political machinery to establish institutions and shape national ideologies. The latter relied heavily on selected perspectives closely tied to Catholicism, an aversion to the British crown, and a strong emphasis on Gaelic culture (including the Gaelic League and Gaelic Athletic Association) (Hopkinson 2004). Publicly acknowledging its geopolitical connection with the British state, especially during WWI, was a delicate matter. For example, W.T. Cosgrave's government permitted public commemorations to honor Irish participation in WWI but did not provide any financial support or attend the events.

Cumann na nGaedheal held majority until 1932 when de Valera's newly formed republican party, Fianna Fáil,[3] emerged as the leading force for a period of Irish

3 de Valera formed Fianna Fáil after he lost electoral leadership of Sinn Féin.

political history. After Cumann na nGaedheal's loss, it merged with additional parties to become pro-treaty Fine Gael in 1933. Thus, the two largest political parties, anti-treaty Fianna Fáil and pro-treaty Fine Gael, reflected the political divisions manifested in the civil war.

With his new party, de Valera succeeded Cosgrave as President of the Free State's Executive Council (1932–1937) and orchestrated a highly selective nation-building program. For example, under de Valera, the Free State became more socially conservative with the Catholic Church as a central force in politics and society. In 1937, the church's privileged position was specifically included in the Constitution.

Women, many of whom had previously participated in the revolution and political movements, were targeted by conservatives as scapegoats for political division in the Free State. They were subsequently removed from jury service, barred from industrial employment (i.e., 1935 Conditions of Employment Act), and relegated to subordinate roles in the 1937 Constitution of the Irish Free State (e.g., Ward 1995). Over time, the government-driven narrative for the Irish nation selectively emphasized: Catholicism, family-focused, male-dominated agrarian society,[4] anti-British sentiment, the Irish language, and nostalgic longing for the west as the island's "cultural heartland" (Graham 1997). This narrow representation excluded Protestants from common perceptions of Irishness, regardless of their political affiliations (Hopkinson 2010).

The IRA also began to change during this time. In 1925, it separated from Sinn Féin's political control to become an independent paramilitary army. Additionally, Fianna Fáil refused to tolerate IRA paramilitary violence. For example, after the IRA detonated several bombs in England in 1938, de Valera approved the execution of six IRA members and banned all protests in their support.

Unlike unionists in Northern Ireland, the Fianna Fáil government began to shun British-related commemorations and Irish WWI veterans were not honored for their service (Brennan 1925). For example, while Irish fought alongside their British counterparts at the Battle of the Somme, this event became associated as a British (Protestant) battle, particularly among loyalists (Braniff, McDowell, and Murphy 2016; Grayson and McGarry 2016). As infantry rifleman William Lynas explained:

> the gallantry of our boys [at the Somme] . . . did not disgrace the name of Ulster or their forefathers' . . . they made a name for Ulster that will never die in the annals of history. No doubt Belfast and the rest of Ulster is in mourning today for our boys, who died doing their duty for King and country.
>
> *(Lynas 1916)*

4 With less heavy industry and a more agricultural focus in the south, the Free State remained a primarily rural society with a weak economy dependent on British trade and monetary system.

In contrast, the nascent Irish government carefully constructed Irish national myths that omitted involvement in WWI. Instead, the nation-building projects it promoted framed national identity as invariably distinctly separate from Britain (e.g., Dempsey 2018). This complex and contested form of subjective remembering and forgetting of WWI is discussed in Chapter 8.

de Valera was particularly influential during these early government-driven efforts to shape Irish national memory. He effectively banned formal collaborations with Britain and abolished the oath to the crown. He refused to honor land annuity payments to the state or recognize and commemorate Irish involvement in WWI. In addition, beginning in 1932, he began to dismantle the Anglo-Irish Treaty and attempted to undermine the partition of Ireland. For example, the 1937 Constitution lays claim to the *entire* island of Ireland, defined as Irish "national territory," including a nominal claim to Northern Ireland. While the Free State lacked the jurisdiction to exercise this unfounded claim, it only removed the statement from the Irish Constitution as part of the 1998 peace agreement. The PG later established an independent foreign policy that permitted its declaration of neutrality during World War II (WWII). Its reluctance to join forces with the UK underpinned its policy of neutrality, as well as its desire to avoid joining a war effort 16 years after concluding a civil war (Ranelagh 2012).

A republic realized

On Easter 1949, the Irish Free State left the British Commonwealth to become the Republic of Ireland. Its legislature (the Oireachtas) consists of the President and two houses: the Dáil Éireann (lower house) and the Seanad Éireann (upper house). The Taoiseach, or "chief," is Prime Minister and head of the government. The Republic's early leadership, conservative in its cultural policies and constitutional directives, promoted Irish as "the national language" and Irish sports as an acceptable form of cultural nationalism (Cronin 1998). Irish politics developed a predominantly two-party system between Fianna Fáil and Finn Gael, despite a lack of notable doctrinal differences distinguishing the organizations.

After years of high inflation and rationing, by the 1950s the Republic remained dependent on the British economy and had high emigration rates, particularly among young, single women who left in search of employment opportunities (McLaughlin 1993). When Séan LeMass succeeded de Valera as Taoiseach of the Republic in 1959, he implemented various measures to help stimulate the Irish economy, including the creation of the airline, Aer Lingus, the Irish Sugar Company, and several housing and educational programs. His government also offered tax-free terms to foreign investors and implemented strong export-oriented policies that were particularly attractive after the Republic joined the European Economic Community (EEC, a predecessor of the EU) in 1973.

The Republic's inclusion in the supranational organization (EEC) encouraged socio-political changes. For example, Irish culture became less insular and socially conservative, particularly regarding the role of women. Subsequently, many chose to join the public workforce (McLaughlin 1993). The republican movement in the Republic became more moderate and "mainstream" compared to its Northern Irish counterpart, a distinction forged as a legacy of partition. The association between Catholicism and Irish national identity also weakened in the Republic in the 1990s as part of a movement that the church labeled "Post-Catholic Ireland" (Ganiel 2016).

In May 1998, the Republic held two referendums that signaled key turning points for Irish geopolitics and national identities. In the first, the public voted to adopt the EU currency, the Euro. In addition to the financial and travel possibilities that a common currency facilitates, it also serves as a symbol of collective European identity. More specifically:

> a common currency is one of the most visible "identity markers" that shapes the EU as a taken-for-granted social fact and helps in building an imagined European community.
>
> *(Risse 2003, 489)*

In the second referendum, the public ratified the Good Friday/Belfast Peace Agreement as part of the peace process in Northern Ireland. The geopolitical and societal impacts of this agreement, and the Republic's role in Brexit negotiations are discussed in Chapters 7, 8, and the book's conclusion.

In a state that once barred women from jury service and industrial employment, Mary Robinson helped redefine gender roles in Irish statecraft when she became the first female President of the Republic in 1990. She was succeeded by President Mary McAleese in 1997. In 2017, Leo Varadkar became Taoiseach of the Republic. This Dublin-born Taoiseach, whose father was born in India, became Ireland's first and the world's fourth openly gay head of government. His appointment signaled societal changes in a country once less ethnically diverse and constitutionally tied to the conservative Catholic Church.

The Republic continues to demonstrate the possibilities of small state diplomacy. Building on its anti-colonial roots, it joined peacekeeping missions in Africa in the 1960s. President Mary Robinson served as the United Nation's (UN) High Commissioner for Human Rights (1997–2002). In 1999, the Irish Republic established a formal relationship with the North Atlantic Treaty Organization for peace-support operations. It also actively contributes as an EU member state, including providing leadership to serve as the European Commissioner for Financial Stability, Financial Services, and the Capital Markets Union since 2020. Despite its small size, its role in supranational organizations and international contributions provided a geopolitical foundation that prepared the state when it joined the UN Security Council in 2021.

Northern Ireland

Northern Ireland's exceptional nature of governance underpinned its unique internal dynamic within the British context. The devolved parliament at Stormont had the power to make its own laws for the region,[5] but could not repeal any "Imperial Matters" established by the crown. The Royal Ulster Constabulary (RUC), the region's police force, consisted primarily of Protestant unionists. It served the dual role of maintaining order within Northern Ireland and defending the border with the Republic of Ireland. Cabinet ministers actively or indirectly supported sectarian discrimination in public and private sectors while the minority (Catholics/nationalist/republicans) lacked viable prospects to challenge the political *status quo* in Northern Ireland (e.g., McAuley 2016). Stormont's abolition of Proportional Representation (PR) in the 1920s further exacerbated the institutional and structural discrimination against Catholics by limiting the "one person, one vote" eligibility to "resident occupiers" (homeowners), thus denying most Catholics' (renters) ability to vote. It also established exclusionary and sectarian employment and housing practices that favored Protestant unionists (McKittrick and McVea 2002).

Four significant political parties in Northern Ireland include the Ulster Unionist Party (UUP), Democratic Unionist Party (DUP), Social Democratic and Labour Party (SDLP), and Sinn Féin (SF). Before the 1970s, the Protestant unionist UUP controlled the region's devolved parliament at Stormont. Most UUP cabinet ministers were also members of the Orange Order. Named after William of Orange, this fraternal organization endeavors to maintain Protestant control of the region. The UUP was responsible for widespread institutional discrimination against Catholics (McKittrick and McVea 2002). As James Craig stated while addressing Stormont in 1934, "All I boast is that we have a Protestant parliament and a Protestant state." The more militant unionist and loyalist DUP, founded in 1971 by Protestant preacher Ian Paisley, purports the "racial superiority of Protestants" over Catholics (Sales 1997). The DUP openly challenged proposals for an Irish backstop in 2020 and voted against Brexit withdrawal agreements presented by former PM Theresa May's government and that of Boris Johnson (discussed in the book's conclusion).

In contrast, the SDLP is the region's leading nationalist party. Established in 1970 by John Hume, a central figure in the peace process, the SDLP membership includes Catholics and Protestants. SDLP was more prominent among nationalists and republicans in the 1970s and 1980s until SF gained more "mainstream" public support in the 1990s. More radical than the SDLP, SF initially defended republican paramilitary violence until SDLP leadership helped convince SF to modify its political strategy for peace in the region.

In comparison to the Free State/Republic's weaker agrarian economy, Northern Ireland possessed a more advanced industrial economy (e.g., textiles, engineering, ship construction) and emphasized trade within Britain instead of fostering intra-Ireland trade. These disparate economies further distinguished these two Irish political entities and contributed to complexities surrounding discussions for

5 Northern Ireland controls 12 seats in the British Westminster Parliament in London. Its regional parliament decides local matters including law, health, education, housing, and policing. The British state determines foreign policy, trade, and state defense for Northern Ireland.

a united Ireland (e.g., the Republic of Ireland's Euro monetary system versus the British Sterling Pound).

Identification(s) in Northern Ireland

As discussed in Chapter 1, despite great diversity and exceptions, there are four main categories in which the inhabitants of Northern Ireland are stereotypically classified: unionist and loyalists and nationalists and republicans. While both unionists and loyalists want Northern Ireland to remain in the UK, loyalists are considered the more hardline or extreme of the two classifications. They are also more focused on their individuality within Britain (e.g., McAuley 2016). In contrast, nationalists and republicans want all of Ireland to be reunited as an Irish republic, with republicans being the more hardline/extreme of the two groups (e.g., Tonge and Gomez 2015). Additionally, while Catholic or Protestant religious categorizations may have historically served as boundary markers, increasingly these labels correspond with a political affiliation or an ethnonational heritage.

However, the socio-political relationship among many unionists/loyalists and nationalists/republicans in Northern Ireland with their "parent" states is dynamic and complex. For example, despite unionists' allegiance to the UK, they may be "disdained by their British fellow-citizens, who tend to view unionists and nationalists as 'equally alien'" (Barry 2003, 194). In 2021, British PM Boris Johnson hinted at a willingness to "give up" Northern Ireland if the Irish border continues to be a geopolitically complicated matter (e.g., McKay 2021).

Nationalists, on the other hand, while strongly identifying as "culturally Irish, have felt a political estrangement from their 'parent' state and the main parties in the Irish Republic" (Tonge and Gomez 2015, 281). In an interview, one of my colleagues from Northern Ireland explained:

> Often, I feel as if Republic forgot there are Irish living in the north. Maybe it was the Troubles that made it so hard for those in the south to remember us living up here. The Irish here are in every way just as Irish as the south, but it's like we have become invisible to those in the Republic.
>
> *(personal interview in Coleraine, 2018)*

Geographic distribution and sectarianism

The geographic distribution of the Protestant/unionists in Northern Ireland is concentrated in the eastern part of the region. In contrast, Catholic/nationalists are centered in the west (Tyrone and Fermanagh) and several neighborhoods in north and west Belfast. While Protestants historically outnumbered Catholics within Northern Ireland by almost a 2:1 ratio, the 2011 report and predictions for the 2021 census revealed a growth in the Catholic population (i.e., 40.8% Catholic to

41.6% Protestant). While Northern Ireland accepts immigrants annually, their rate is notably smaller than other areas of the UK and other EU member states, and immigration has a "marginal numeral effect" on the population distribution as a whole (Central Statistics Office 2016).

While those labeled as Catholics only represented 35% of the population of Northern Ireland at its formation, since the seventeenth century, the foremost perception among many within the Protestant community was that of a "siege mentality" as a statistical minority on the island (Nash 2013). In addition, the IRA's attempt to overthrow the regional government (the "Border Campaign" 1956–1962) and the manner in which the Irish Free State's government framed Irish identity as "Catholic" and "anti-British" contributed to Ulster Protestant unionists' increasingly sectarian measures used against Catholics in Northern Ireland as the "enemy within" (Dáil Debates, 19 Sept. 1922, col. 697). Eventually, Catholics in the north turned their back on the Stormont government and hoped for a rapid reunification with the Free State. They began to socially entrench themselves within their own cultural clubs, religious organizations as well as their own sectarian-divided schools and neighborhoods.

Some of the economically powerful in Northern Ireland utilized sectarianism as a manipulation tactic. For example, when Belfast labor organizers attempted to unite working-class Protestants and Catholics to demand better working conditions, industrialists and the social gentry of the city avoided a unified, class-based campaign by increasing their support of the sectarian Orange Order fraternal lodges. These lodges employed discriminatory methods to ensure unionist/loyalist dominance over employment sectors in Northern Ireland. Their methods ultimately dismantled efforts to unify Belfast's working-class across sectarian lines and the economic elite continued to encourage sectarianism "to preserve its prerogatives by fostering sectarian strife among its opponents [labor unions and organizers] while Protestant workers ignored their real economic interests and gave their support to the Unionist party and Orange Order" (Henderson, Lebow, and Stoessinger 1974, 209).

The competing perceptions of place and belonging entrenched ethnonational divisions. In the 1960s, civil rights advocates in Northern Ireland, inspired by similar efforts led by the African American community in the US, began to campaign for an end to the institutionalized and social discrimination against the Catholic community. When reactions to their campaigning became violent, the region entered the Troubles, a violent period of protracted conflict (discussed in subsequent chapters).

Despite their great contributions to society, peace advocacy, and even paramilitary organizations during the Troubles, women in Northern Ireland faced widespread prejudice (McDowell 2008). Commonly framed as "guardians" of the family and moral values, socially constructed expectations for women focused on their reproductive "obligations." For example, Irish women were encouraged to have many sons to fight for Ireland's liberation, while certain traditional interpretations of Protestantism expected wives to self-sacrifice all for family and husband (e.g.,

Sales 1997). Most were not encouraged to work outside of the home, and if they did, women faced widespread economic discrimination in Northern Ireland. As women began to fill leadership roles, particularly during the Troubles, many of their efforts were overlooked in media reports and historical accounts (McDowell 2008). Geographer Lorraine Dowler investigated gender in times of conflict and revealed how women in Northern Ireland were historically "marginalized in the consciousness of those who researched the events of war" (159). In contrast, Dowler demonstrated that "both men and women were equally joined in the resistance in Northern Ireland" (1998, 173).

Legacies of the border and divisions

Undeniably, the border in Ireland continues to be a significant factor in inter-Irish, British–Irish, and recently British–Irish–EU relations. During the Troubles, many nationalists/republicans perceived the border as an artificial divide, whereas unionists/loyalists believed it necessary to demarcate the UK's Northern Ireland from the Republic. Due to the increasing violence in the region, the Irish border region was increasingly militarized and "hardened." Cross-border driving routes were blockaded by steel girders, ground "spikes," and guarded from watchtowers and by the military on most frontier roads. Subsequently, the border region suffered from under-investment and decreased population (Hayward 2018). Crossing the Irish border became more physically and socially difficult, underpinning perceptions of difference and "foreign" along the border (e.g., MacDonald 2018). As one interviewee who lived in a frontier town in Cavan explained:

> The border divided my family from my uncle's farm on the other side of the border. As a kid, I noticed the shops in the north sold different foods, like Mars Bars, which you couldn't get in the south . . . Then, during the Troubles, I remember the military checkpoints that stopped us at the border when we drove north. I also remember the helicopters and British squaddies patrolling the border area . . . I knew people who lived on the other side of the border. We were all "borderers," people who crossed the border for family, work, or shopping, but the government categorized and treated us differently, according to our passports and religion.
>
> *(personal interview, 2018)*

Indeed, physical and societal bordering processes impacted inhabitants of Ireland, but perhaps most significantly individuals living along the border region.

Eventually, intergovernmental agreements began to alter the significance and functionality of the border. For example, the 1985 Anglo-Irish Agreement (discussed in Chapter 4) created a political mechanism through which the Republic could express opinion(s) pertaining to Northern Ireland, and Westminster promised a "determined effort" to address any concerns raised (Hayward 2018). More significantly, EU integration and the 1998 peace agreement (discussed in

Chapter 4) encouraged permeability and invisibility of the border for a shared "common ground." Additionally, the EU's Single Market (1993) removed custom barriers and fostered free movement of capital, goods, services, and people in Ireland. Within the context of European integration and the peace agreement, the border transformed from a dividing line into more of a symbolic site of potential for cooperation, integration, and community-building.

The 1998 peace agreement also provided support for new cross-border organizations that emphasized trade and cooperation between the Republic and Northern Ireland. As the satirical Twitter account @BorderIrish, which speaks as the voice of the Irish border, explained:

> The Good Friday Agreement was a miracle of a kind. A negotiation, a compromise, a realpolitik solution where none seemed possible . . . I have become a functioning, actually-existing constructive ambiguity, an accommodation of irreconcilabilities . . . my invisibility is perfectly poised between two political ideologies – one can pretend I'm not there and the other can pretend I am, and both can think they're right. Genius! I've persuaded everyone that I am (a) invisible and (b) crucial to the maintenance of the fragile consensus . . . I'm a post-borderist border who wants to stay post-borderist, thank you very much. That annoys people who want firm lines, certainly and absolutes, and things that are singularly simplesimplesimple, but I can't be that. I won't.
>
> *(2019, 11, 67)*

While the EU and peace agreements reduced some of the functional significance of the border, its presence fosters distinctions between the two societies, cultures, and selected national narratives. Despite shared histories, cultures, and identities, the border divides individuals between two distinct political units. In Northern Ireland, the production and institutionalization of sectarian discrimination exacerbated and, in many cases, created divisions within its society. Its legacy and that of the Troubles persists in the minds of many in Northern Ireland and is transmitted to new generations through partisan narratives, mental maps, sectarian commemorations, divided schools, and separate institutions (e.g., McAuley 2016; McDowell and Shirlow 2011). Chapters 4–7 explore the role that these situated ethnonational geographies play in the production of division and difference as well as peace and reconciliatory work within Northern Ireland.

References

Barry, John. 2003. "National Identities, Historical Narratives and Patron States in Northern Ireland'." In *Political Loyalty and the Nation-State*, 189–205. Oxfordshire: Taylor and Francis.

Belton, A. 1922. *Correspondence to Lord Midleton, 3 October*. London: Midleton Papers.

Bielenberg, Andy. 2013. "Exodus: The Emigration of Southern Irish Protestants during the Irish War of Independence and the Civil War." *Past & Present* 2018.

Borgonovo, John. 2011. *The Battle for Cork, July – August 1922*. Cork: Mercier Press.

Brambilla, Chiara, and Reece Jones. 2020. "Rethinking Borders, Violence, and Conflict: From Sovereign Power to Borderscapes as Sites of Struggles." *Environment and Planning D: Society and Space* 38 (2): 287–305.

Braniff, Máire, Sara McDowell, and Joanne Murphy. 2016. "Editorial Introduction." *Irish Political Studies* 31 (1): 1–3. https://doi.org/10.1080/07907184.2015.1126923.

Brennan, Joseph. 1925. "Joseph Brennan Papers." Manuscripts Reading Room. National Library of Ireland.

Central Statistics Office. 2016. "Irish Census Report." https://www.cso.ie/en/media/csoie/newsevents/documents/pressreleases/2017/prCensussummarypart1.pdf

Collins, Michael. 1922. "Collins to Louis J. Walsh, Derry, 1 February 1922." Department of An Taoiseach. National Archives of Ireland.

Cronin, Mike. 1998. "Enshrined in Blood the Naming of Gaelic Athletic Association Grounds and Clubs." *The Sports Historian* 18 (1): 90–104. https://doi.org/10.1080/17460269809444771.

Dáil Debates. 1922. *1922*. Dublin: Oireachtas.

Dempsey, Kara E. 2018. "Creating a Place for the Nation in Dublin." *The City as Power: Urban Space, Place, and National Identity* 27.

———. 2022. "Fostering Grassroots Civic Nationalism in an Ethno-Nationally Divided Community in Northern Ireland." *Geopolitics*: 1–17. https://doi.org/10.1080/14650045.2020.1727449.

Dodds, Klaus. 2013. "'I'm Still Not Crossing That': Borders, Dispossession, and Sovereignty in *Frozen River* (2008)." *Geopolitics* 18 (3): 560–83. https://doi.org/10.1080/14650045.2012.749243.

Dowler, Lorraine. 1998. "'And They Think I'm Just a Nice Old Lady' Women and War in Belfast, Northern Ireland." *Gender, Place & Culture* 5 (2): 159–76. https://doi.org/10.1080/09663699825269.

Ferriter, Diarmaid. 2015. "Hearts of Stone in Ireland's Civil War." *The Irish Times*, March 7, 2015. www.irishtimes.com/culture/heritage/hearts-of-stone-in-ireland-s-civil-war-1.2125800.

Foster, Gavin M. 2015. *The Irish Civil War and Society: Politics, Class, and Conflict*. London: Palgrave Macmillan.

Ganiel, Gladys. 2016. *Transforming Post-Catholic Ireland: Religious Practice in Late Modernity*. Oxford: Oxford University Press.

Graham, Brian. 1997. *In Search of Ireland: A Cultural Geography*. East Sussex: Psychology Press.

Grayson, Richard S., and Fearghal McGarry. 2016. *Remembering 1916: The Easter Rising, the Somme and the Politics of Memory in Ireland*. Cambridge: Cambridge University Press.

Hayward, Katy. 2018. "Brexiting Borderlands: The Vulnerabilities of the Irish Peace Process." *Accord: An International Review of Peace Initiatives* 4: 78–80.

Heintz, Matthew. 2009. "The Freedom to Achieve Freedom: Negotiating the Anglo-Irish Treaty." *Intersections* 10 (1): 431–51.

Henderson, Gregory, Richard N. Lebow, and John G. Stoessinger. 1974. *Divided Nations in a Divided World*. New York, NY: David McKay Company.

Hopkinson, Michael. 2004. *The Irish War of Independence*. Montreal: McGill-Queen's Press-MQUP.

———. 2010. *Green Against Green: The Irish Civil War*. 2nd ed. Dublin: Gill & Macmillan.

Johnson, Corey, Reece Jones, Anssi Paasi, Louise Amoore, Alison Mountz, Mark Salter, and Chris Rumford. 2011. "Interventions on Rethinking 'the Border' in Border Studies." *Political Geography* 30 (2): 61–69.

Kissane, Bill. 2005. *The Politics of the Irish Civil War.* Oxford: Oxford University Press.
Komarova, Milena, and Katy Hayward. 2019. "The Irish Border as a European Union Frontier: The Implications for Managing Mobility and Conflict." *Geopolitics* 24 (3): 541–64.
Lee, Joseph. 1990. *Ireland, 1912–1985: Politics and Society.* Cambridge: Cambridge University Press.
Lynas, William. Letter. 1916. "Personal Letter to His Wife, Written by Rifleman William Lynas, 15th Royal Irish Rifles." *Imperial War Museum*, July 15, 1916.
MacDonald, Darach. 2018. *Hard Border.* Dublin: New Island Books.
"MacMahon at 1924 Army Enquiry." 1924. Dublin: University College Dublin Archives.
McAuley, Jim W. 2016. *Very British Rebels? The Culture and Politics of Ulster Loyalism.* London: Bloomsbury.
McCoole, Sinead. 2004. *No Ordinary Women: Irish Female Activists in the Revolutionary Years.* Dublin: University of Wisconsin Press.
McDowell, Sara. 2008. "Commemorating Dead 'Men': Gendering the Past and Present in Post-Conflict Northern Ireland." *Gender, Place & Culture* 15 (4): 335–54. https://doi.org/10.1080/09663690802155546.
McDowell, Sara, and Peter Shirlow. 2011. "Geographies of Conflict and Post-Conflict in Northern Ireland: Conflict in Northern Ireland." *Geography Compass* 5 (9): 700–709. https://doi.org/10.1111/j.1749-8198.2011.00444.x.
McKay, Susan. 2021. "Northern Ireland Is Coming to an End." *The New York Times*, 2021. www.nytimes.com/2021/06/30/opinion/northern-ireland-centenary.html.
McKittrick, David, and David McVea. 2002. *Making Sense of the Troubles: A History of the Northern Ireland Conflict.* Chicago: New Amsterdam Books.
McLaughlin, Eugene. 1993. "Ireland: From Catholic Corporatism to Social Partnership." In *Comparing Welfare States: Britain in International Context*, edited by Allan Cochrane and John Clarke, 223–60. London: Sage.
Moody, Theodore William, and Francis X. Martin. 2001. *The Course of Irish History.* Cork: Mercier Press.
Nash, Catherine, and Bryonie Reid. 2013. *Partitioned Lives: The Irish Borderlands.* 1st ed. New York, NY: Routledge.
O'Leary, Brendan, and John McGarry. 2016. *The Politics of Antagonism: Understanding Northern Ireland.* London: Bloomsbury Publishing.
Ranelagh, John. 2012. *A Short History of Ireland.* Cambridge: Cambridge University Press.
Rankin, K. J. 2007. "Deducing Rationales and Political Tactics in the Partitioning of Ireland, 1912–1925." *Political Geography* 26 (8).
Risse, Thomas. 2003. "The Euro between National and European Identity." *Journal of European Public Policy* 10 (4): 487–505.
Sales, Rosemary. 1997. *Women Divided: Gender, Religion and Politics in Northern Ireland.* East Sussex: Psychology Press.
Shuttleworth, Ian. 2007. "Reconceptualising Local Labour Markets in the Context of Cross-Border and Transnational Labour Flows: The Irish Example." *Political Geography* 26 (8): 968–81.
Tonge, Jonathan, and Raul Gomez. 2015. "Shared Identity and the End of Conflict? How Far Has a Common Sense of 'Northern Irishness' Replaced British or Irish Allegiances since the 1998 Good Friday Agreement?" *Irish Political Studies* 30 (2): 276–98.
Ward, Margaret. 1995. *In Their Own Voice: Women and Irish Nationalism.* New York, NY: Atrium.
Waterman, Stanley. 1987. "Partitioned States." *Political Geography Quarterly* 6 (2): 151–70.

4

"THE TROUBLES" ACROSS VARIOUS SCALES

In the late 1960s, competing claims to territory and power in Northern Ireland erupted into a tumultuous period known as "the Troubles." Violence during the Troubles was highly territorialized and fueled by unrest over the constitutional status of Northern Ireland, the presence of paramilitaries and security forces, and systematic sectarian discrimination (e.g., McDowell and Shirlow 2011). When the RUC police was unable to quell widespread rioting and violence, the British state deployed thousands of troops,[1] armored carriers, and amphibious assault vessels to the region, in an effort known as Operation Banner (1969–2007).

Initially, most welcomed the military as a protective force from sectarian attacks. However, many Catholics/nationalists quickly perceived the army as an extension of unionist control in Northern Ireland. After increased military and police residential searches, loyalist attacks, and curfews enforced in nationalist/republican/Catholic enclaves,[2] these neighborhoods self-barricaded to impede armed forces from entering. Residents created "No-Go areas" by blocking the streets on the external edges of their neighborhoods with burnt-out cars and public buses. The military, in turn, targeted these nationalist/republican/

1 In addition to the army, the Ulster Special Constabulary ("B-Specials") assisted the RUC, but this all-Protestant militia exacerbated Catholic distrust and was eventually disbanded. The state deployed the Special Air Service (SAS) and Security Service (MI5) to Northern Ireland during Operation Banner. Certain SAS forces were accused of collusion with loyalist paramilitary organizations and "shoot to kill" tactics in Northern Ireland.

2 For example, the Falls Road Curfew (July 3–5, 1970) is considered the "final poisoning of the initially good relations between Catholics and British troops" (McKittrick and McVea 2002, 61).

DOI: 10.4324/9781003141167-4

Catholic "No-Go areas." In an interview, one Belfast resident recalled the experience:

> It was the scariest thing. It was like the Falls [Catholic/republican enclave] had been invaded. Military troops patrolled the streets with machine guns, like we were in a prisoner of war camp, you know? I remember walking to school or to the shops, we'd pass British armed soldiers in the streets everywhere.
>
> *(personal interview in Belfast, 2016)*

During the Troubles, security forces and paramilitary organizations (republican and loyalist) claimed the lives of over 3,500 and injured 40–50,000 people in a region of only 1.5 million inhabitants (Mesev, Shirlow, and Downs 2009). If one extrapolates Northern Ireland's death toll to the population of the United States, it would yield over 526,000 deaths – a sum greater than the total number of Americans who died in WWII (Hayes and McAllister 2001).

To geopolitically contextualize the interviewee's statement, the rise in paramilitary violence, and the increased militarization of Northern Ireland, this chapter examines the spatial nature of political struggles and peace efforts during this period. This chapter is not intended to provide a comprehensive history of the Troubles. Instead, it offers a multiscalar investigation (i.e., regional, state, and international)[3] of spatialized experiences, geopolitical narratives, and significant legislation during this tumultuous era. Political geographers conceptualize scale as socially constructed, dynamic, and overlapping "forms of hierarchy" (e.g., Marston and Neil 2001; Paasi 2004; Dempsey 2016). Some suggest that political realities do not reside solely on one scale. Instead, as scales interconnect, actors across multiple scales may influence an action or event (e.g., Barrett 2013).

A scalar investigation provides a "geographic understanding of context" in which "situation and dynamism can be most fruitfully examined" (Flint 2016, 5–6). Thus, analyzing the Troubles through a multiscalar lens fosters greater understanding of the networked forces that contributed to complex geographies of conflict and peace efforts in Northern Ireland. It also offers insight into territorialized identities, uneven power relationships, and the production of violence and discrimination in place-specific contexts.

A brief history of the Troubles

To contextualize the multiscalar discussion explored throughout this chapter, the following section provides a brief history of the Troubles in Northern Ireland (see Figure 4.1). While the timeline of the Troubles is commonly framed between 1969

3 Chapters 5, 6, and 7 examine conflict and peace efforts at the local or "neighborhood" scale.

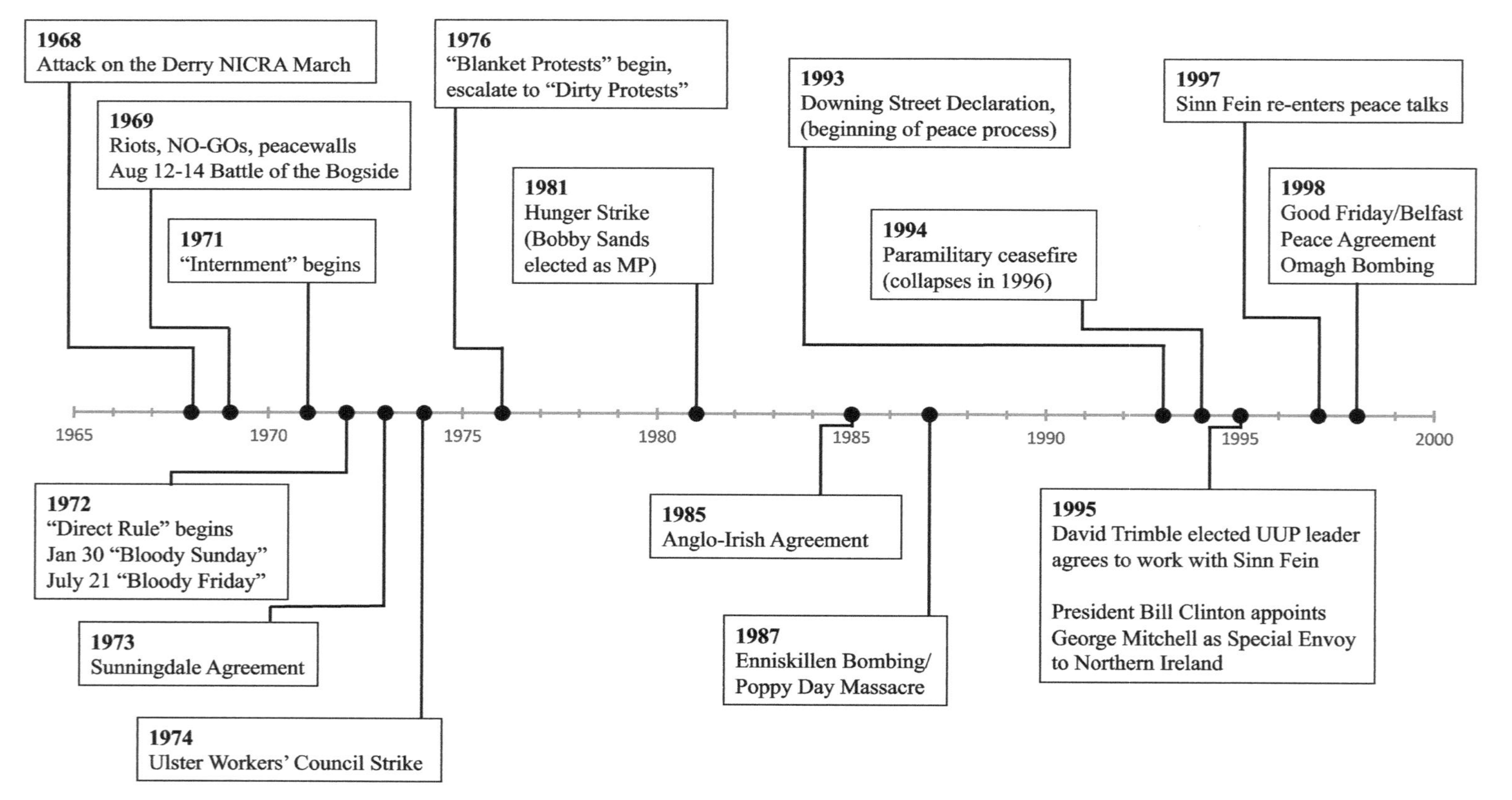

FIGURE 4.1 Some of the significant events during the Troubles.

and 1998, many identify the onset of this conflict as a result of a violent attack on a civil rights march that transpired on October 5, 1968, in Londonderry/Derry.[4]

When the British government established Northern Ireland and discretely transferred governance to unionists, the newly empowered were a minority on the island; Protestants represented 1/5 of the island of Ireland (Sales 1997). Trying to bolster their presence in Ulster, unionists excluded Catholics from all major political appointments and forged institutional discrimination against its Catholic citizens. The devolved Northern Irish parliament in Stormont was responsible for key domestic matters including housing, health, education, law, and local policing. Subsequently, the Ulster Unionist Party (UUP) was able to control Stormont for over 50 years. As the first Prime Minister of Northern Ireland (1921–1940), Lord Craigavon stated, "All I boast is that we have a Protestant parliament and a Protestant state" (quoted in Farrell 1980, 92).

To confront institutionalized regional discrimination, the Northern Ireland Civil Rights Association (NICRA) was formed in 1967. The NICRA called for a peaceful end to discrimination against Catholics and began organizing marches and sit-ins. When rioters and some local law enforcement attacked NICRA demonstrators, it evoked international condemnation against the region's discriminatory laws and police actions. Growing external disapproval pressured the Stormont government to enact some regional reforms, but most were symbolic in nature and fell short of the civil rights advocates' fundamental demands (Sales 1997). While most nationalists/republicans considered the reforms inadequate, many unionists/loyalists believed them too drastic. This belief contributed to further escalating tensions within Northern Ireland; and ultimately Northern Ireland's PM, Terry O'Neill (1963–1969), stepped down.

In 1969, growing internal sectarian tensions erupted into riots and widespread violence throughout Northern Ireland. When the RUC was unable to maintain control, the British Army attempted to restore order through "Operation Banner." At the culmination of Operation Banner's first year, sectarian tensions continued to rage and the presence of barrier walls or "peacelines," "No-Go" areas (i.e., self-barricaded areas to impede the RUC and loyalist rioters), and contested areas became highly segregated along sectarian lines. That same year the IRA launched their guerrilla-style "Long War" against the government (e.g., Moloney 2002).

The British army was unable to establish order as tensions escalated, which further fomented paramilitary violence and the formation of radicalized political parties in Northern Ireland. In response, in 1971 the government introduced the permitted imprisonment without a trial (i.e., "Internment"). Northern Ireland's PM, Brian Faulkner (1971–1972), supported Internment because he believed he was "at war with terrorists, and in a state of war many sacrifices have to be made." As the political situation in Northern Ireland continued to deteriorate, the region's unionist-controlled parliament was suspended, dissolved, and superseded by the

4 Unionists and loyalists commonly refer to this city as Londonderry, while nationalist and republicans use its pre-colonial name, Derry.

London-based British government's "Direct Rule" in 1972 in an effort to re-establish law and order in the region.

1972 was the most violent year (i.e., number of deaths) of the conflict. For example, on January 30, 1972, a Parachute Regiment of the British Army killed 13 unarmed civil rights marchers in Londonderry/Derry. The violence of the events of "Bloody Sunday" was captured on film and the footage generated worldwide outrage and a judicial inquiry by the British government. The unprovoked attack garnished multiscalar support for the IRA and increased enlistment in its ranks throughout the island (Moloney 2002). In an effort to quell the mounting violence, Northern Ireland's Secretary of State began secret talks with the IRA to discuss potential routes to peace. When these efforts proved unsuccessful, the IRA launched a violent campaign to disrupt daily life throughout the region. One of their most notorious bombing events, known as "Bloody Friday," killed nine and injured 130 civilians in 1972. The widespread carnage prompted the British Army to launch a region-wide military operation in Northern Ireland known as "Operation Motorman," which targeted the IRA and "No-Go" areas, becoming the largest British military campaign since the Suez Crisis. Despite extensive British military presence throughout the region, the IRA actively continued their bombing campaign in Northern Ireland. They also expanded their attacks into England, killing over 200 people and assassinating key military or diplomatic targets throughout Europe.

Institutionalized peace efforts began in 1973 but proved unsuccessful until 1998. The Sunningdale Agreement (1973) was part of an effort to restore self-government in the region and proposed a power-sharing, devolved administration. It also permitted the Republic of Ireland an advisory role in internal matters in Northern Ireland, which angered the loyalist/unionist community. Their protests culminated with a loyalist-organized regional Ulster Workers Council Strike in 1974 that contributed to the collapse of the Sunningdale government.

The conflict continued after the failure of the Sunningdale Agreement. In the milieu of the tumult, the 1981 republican prisoners' hunger strike became one of the most notable events during the Troubles. In an effort to gain "special category status," which would have distinguished the protestors as political prisoners instead of criminals, participants launched a hunger strike in Northern Ireland's Long Kesh "Maze" prison. Ten inmates, including Bobby Sands, who was elected as a British MP during this protest, died as a result of the strike. The death of the protestors sparked riots, generated international sympathy for the republican cause, and encouraged republican Sinn Féin to engage in mainstream British politics. Prior to this time, its members refused to take their elected seats in the British parliament as a political demonstration of their rejection of British rule in Ireland.

In 1985, the Anglo-Irish Agreement (AIA) laid the foundational groundwork for peace in Northern Ireland. One of its central tenets included the "consent principle" assuring that the region could only be dissolved with the consent of a majority of its inhabitants. While the AIA gained support from moderate parties, the political extremes in Northern Ireland staunchly opposed it. The IRA continued its bombing campaign, including the Enniskillen Bombing ("Poppy Massacre") during a WWI remembrance ceremony in 1987. The bombing claimed the lives of 11 (ten were civilians) and injured 63 others, including women and children who had gathered to commemorate those who had fallen in the war.

At the culmination of this tumultuous decade, secret talks among the British government, the nonviolent nationalist Social Democratic and Labour Party (SDLP), Roman Catholic clergy, and Sinn Féin focused on strategies to terminate the violence in Northern Ireland. In 1993, the Irish and British governments established the Downing Street Declaration (DSD). While the loyalist DUP condemned the DSD, the more mainstream Official Unionists, Irish nationalists, and much of Sinn Féin's leadership cautiously supported it.

The following year, the IRA began their ceasefire (without decommission) in August, which preceded the reciprocal armistice by their loyalist (UVF and UDA) counterparts. Subsequently, Sinn Féin representative Martin McGuiness was invited to Stormont for exploratory talks. This encouraged the Republic of Ireland to establish the "Forum for Peace and Reconciliation" to foster multiscalar peace negotiations to progress. In 1995, David Trimble became the leader of the UUP and eventually agreed to negotiate with republican Sinn Féin. He and SDLP leader, John Hume, would jointly receive the Nobel Peace Prize in recognition of their cooperative peace efforts (McKittrick and McVea 2002).

The United States provided significant support for peace efforts in Northern Ireland. In 1995, President Bill Clinton appointed US senator George Mitchell to chair peace talks that would ultimately culminate with the 1998 Good Friday/Belfast Agreement. Backed by support of moderates as well as representatives of loyalist and republican political parties, this power-sharing agreement represented a seismic shift in Northern Irish politics. It established the Northern Irish Assembly, Northern Ireland's devolved legislature within the UK, and fostered a fragile peace amidst a legacy of violence in Northern Ireland.

The regional scale – Northern Ireland

During the Troubles, there were significant socio-political disparities unique to Northern Ireland within the UK. In response, civil rights demonstrations protested the institutionalized inequality underpinned by the perceptions among "old-style unionists" of Catholics as the "enemy within" (Barry 2003). This section focuses on the influence of civil rights campaigns and political protests in Northern Ireland during the Troubles. However, this is preceded by a brief discussion of regional political power and armed forces, to provide some context (see Chapter 3 for descriptions of the political parties).

Political power in Northern Ireland

The UUP-controlled regional government institutionally discriminated against Catholics in Northern Ireland. It altered election regulations and failed to implement British equitable employment policies for its Catholic citizens. The resultant discrimination and economic disparity between Protestants and Catholics in Northern Ireland became the largest inequity rate in the UK (Cairns and Darby 1998). Widespread gerrymandering – the practice of forging electoral boundaries around concentrated populations with similar socio-political views into a single area while producing more electoral districts around the minority to reduce the influence of the majority population's vote – assured unionist electoral victories throughout the region. For example, the unionist/Protestant minority (8,700 people) were allotted two districts that totaled 12 seats, while the larger nationalist/Catholic population (14,400) was relegated to a single electoral district with eight seats (McKittrick and McVea 2002).

The regional government also regulated construction and allotment of public housing to prioritize Protestants. When a political candidate's single secretary received housing before numerous Catholic families, widespread protests erupted throughout Northern Ireland. Indeed, residential allocation was so problematic in Northern Ireland that some researchers argue it triggered the onset of the Troubles (Gaffikin et al. 2016). Despite their protests, Catholic families continued to struggle to find housing. If Catholics received housing accommodations, they were geographically concentrated into large, low-quality public residential blocks, sometimes called "housing estates." Military foot soldiers, helicopters, and watchtowers disproportionately monitored these structures (Patterson 2013).

As a former resident of a housing estate in Londonderry/Derry explained:

> Before we got the flat in the estates, my family squatted in an abandoned American WWII ammunition barrack at the edge of town. I lived there with 7 siblings, 2 parents, 1 grandparent and 2 uncles. But once we finally got a flat, the RUC often raided our home at night. They interrogated my brothers about being in the IRA. My brothers weren't in the IRA, but the RUC kept coming back – because of where we were living. Our housing labeled us. Growing up in the estate really impacted my life, my identity, and how I see the world.
>
> *(personal interview in Derry, June 2012)*

Armed forces

Republican paramilitary organizations,[5] loyalist paramilitary organizations,[6] and Government military/security forces (e.g., Operation Banner or the Royal Ulster Constabulary) were the three most significant armed forces active during the Troubles.

5 For example, the Irish Republican Army (IRA, originally PIRA), the Official Irish Republican Army (IRA O), Continuity (CIRA) and Real Irish Republican Army (RIRA), the Irish National Liberation Army (INLA), the Civilian Defence League (CDL), and Direct Action Against Drugs (DAAD, alleged front group for the IRA) (Mesev, Shirlow, and Downs 2009).

6 For example, the Ulster Volunteer Force (UVF), the Ulster Defence Association/Ulster Freedom Fighters (UDA/UFF), the Red Hand Commando (RHC), the Red Hand Defenders (RHD), the Protestant Action Group (PAG), and the Protestant Action Force (PAF).

Paramilitary forces in Northern Ireland

Several paramilitary organizations (e.g., the IRA, INLA and the UVF, UDA, UFF) were active during the Troubles. In fact, some researchers describe the Troubles through two major conflicts: the battle between the IRA and security forces; and loyalist paramilitary violence against Catholic civilians (Gregory et al. 2013). Divisions and infighting among republican as well as loyalist paramilitaries also resulted in a number of deaths, including their own civilian populations caught in interdivisional fighting[7] (McKittrick and McVea 2002).

Republican paramilitaries

At the beginning of the Troubles, the relatively "moribund" IRA[8] re-emerged as a self-declared protective force for Catholics. The smaller socialist-republican paramilitary Irish National Liberation Army (INLA) formed in 1974 to protest the IRA's acceptance of a ceasefire in 1972. The IRA fashioned itself as the "army of the people" to defend Catholics against Protestant violence and British security forces in Northern Ireland. The IRA demanded an end to British control of Ireland through its guerrilla-style war of attrition it called the "Long War" (Patterson 2013).

As part of its "war," the IRA divided into small cells and launched terrorist bombings intended to make British control of Northern Ireland impractical. The IRA believed its campaign was an anti-colonial fight. Indeed, prior to the 9/11 attacks, terrorism was more frequently framed as an anti-imperial or anti-colonial revolutionary struggle that included the IRA's efforts to end the British colonialization of Ireland (Gallaher 2012).

While the IRA targeted security forces and loyalist paramilitaries, its indiscriminate bombing campaign resulted in several civilian deaths. As a Belfast native from loyalist Sandy Row explained:

> During the Troubles, my uncle was killed – murdered – in one of the IRA bombings. The IRA terrorized Belfast and terrorized Northern Ireland. The police never found those responsible for the bomb, so they're walking free to this day. There's been no justice for my uncle, no justice at all.
>
> *(personal interview in Belfast, August 2017)*

7 For example, a loyalist feud between UDA and UVF paramilitaries resulted in several deaths (1974–1975). Republican paramilitaries also had internal conflicts, especially when the IRA split into the IRA Provisionals ("Provs") and IRA Officials in 1969. The IRA further fragmented as members from the PIRA formed the Real IRA (RIRA). The RIRA had defectors that formed the dissident faction Óglaigh na hÉireann (ONH) and the "New IRA," which rivaled a previously formed Continuity IRA (CIRA).

8 The Provisional Irish Republican Army – "Provisionals" (PIRA) split from the Official Irish Republican Army "Officials" (IRA O) and PIRA became the primary republican paramilitary organization (IRA) in Northern Ireland.

Loyalist paramilitaries

In retaliation, loyalist organizations attacked IRA members and Catholic civilians (Hayes and McAllister 2001). For example, the UVF's indiscriminate violence against Catholics paralleled that of the IRA's attacks against Protestants. In 1971, the Ulster Defense Association (UDA) was formed to defend Protestants from IRA attacks while its more radical faction, the Ulster Freedom Fighters (UFF), targeted Catholic civilians and the IRA.

Despite the British military presence, paramilitary control of corresponding sectarian areas continued. Opposing paramilitaries, "internal policing," and even intra-paramilitary fighting was common in sectarian neighborhoods. For example, while local paramilitaries defended local residents from opposing paramilitary attacks, they also "self-policed" their own local residents through punishment beatings against informers or petty criminals (Gregory et al. 2013).

The role of government military/security forces

When the British Army and the RUC failed to stop the escalation of violence in the region, the state implemented "Operation Demetrius" in 1971. Locally known as "Internment," this controversial policy increased policing powers and permitted imprisonment without a trial. As the government apprehended more innocent civilians than paramilitaries (e.g., as portrayed in the 1993 film, *In the Name of the Father*), Internment received international condemnation. The European Court later indicted the British state for this policy, as it fostered torture of detainees through brutality described as:

> an inhuman or savage form of cruelty and that cruelty implies a disposition to inflict suffering, coupled with indifference to or pleasure in the victim's pain.
> *(BSSRS 1974, 31)*

As the political and social environment in Northern Ireland continued to deteriorate, the state dissolved the regional parliament at Stormont and implemented British "Direct Rule" in 1972.

The state-imposed "Direct Rule" had an immense bearing within Northern Ireland. For example, the dissolution of Stormont was highly unpopular among unionists that unenthusiastically accepted this political usurpation. Many feared it would result in the eventual reunification of Ireland. Their anxieties were further exacerbated when many Irish nationalists initially applauded the seizing of unionist-controlled Stormont.

Conversely, the IRA abhorred the increased British military presence in Northern Ireland. It escalated its bombing campaign in an effort to violently disrupt daily life in the region. One of the most notorious events, known as "Bloody Friday" in 1972, resulted in 23 bombs that claimed the lives of nine and injured 130 civilians throughout Belfast.[9] The IRA attacks triggered a reciprocal loyalist paramilitary campaign

9 They expanded into England as well, killing over 200 people, and assassinating key military and diplomatic targets.

that targeted Catholics, nationalists, and republicans throughout the region. When security forces subsequently escalated their campaign against paramilitary organizations in Northern Ireland, it resulted in the Troubles' most violent year, 1972.

Civil rights campaigns

Inspired by Dr. Martin Luther King's nonviolent civil rights marches, the Northern Ireland Civil Rights Association (NICRA) began organizing sit-ins and civil rights protests in 1967. NICRA called for widespread regional reforms, including equal voting rights in local elections, fair allotment of public housing, termination of gerrymandering and employment discrimination, police reform (including disbandment of the Protestant elite police force, the "B specials"), and the end of Internment.[10] Its membership consisted of activists from various religious/ethnonational backgrounds, united by the shared belief in the need for equality and universal civil rights. As civil rights leader Bernadette Devlin explained:

> We [NICRA] refused to accept the politicians' logic that the problems could be seen in terms of Catholic versus Protestant . . . The civil rights movement was interested in people's needs.
>
> *(1969, 58)*

Indeed, as one Protestant NICRA member shared in an interview:

> I was inspired by MP Ivan Cooper. He was a founding member of the SDLP, and like me, a Protestant. I saw him speak once. He spoke about how Internment, gerrymandering, and employment discrimination was crippling Northern Ireland and our hope for a just society . . . He inspired me to stand against the injustice, so I joined NICRA.
>
> *(personal interview, May 2018)*

NICRA was relatively unknown outside of Northern Ireland until the RUC and rioters assaulted civil rights marchers with batons and water cannons on October 5, 1968, in Londonderry/Derry. A television reporter for Irish RTE, the Republic's national television and radio station, released recorded footage of the attack through the public media (McKittrick and McVea 2002). Images of the RUC attacking these peaceful demonstrators sparked international censure for Northern Ireland's discriminatory laws and police actions. Bernadette Devlin, who participated in the demonstration, explained that the attack "was my first realization that the police hated us" (Devlin 1969).

10 In the 1980s, police utilized informants (i.e., "supergrasses") to testify against known associates in exchange for immunity. While judges later challenged the validity and ethical standards of the practice, it resulted in many paramilitary arrests.

In January 1969, NICRA organized a march from Belfast to Londonderry/Derry in a demonstration modeled after Dr. King's march from Selma to Montgomery. However, during NICRA participants' journey, loyalists ambushed the activists. The media widely broadcast these violent attacks, which ignited sectarianism throughout the region. For example, the "Battle of the Bogside" transpired in the "Bogside" republican enclave in Derry (August 12–14, 1969) when residents returning from the Belfast-Derry NICRA march were harassed by loyalist paramilitaries and police. In response, residents created a "No-Go" area that provoked further loyalist and police aggression. Riots subsequently erupted in Belfast in response to the turmoil (discussed in Chapter 5).

While NICRA functioned at the regional scale, it began to appeal for international support for their demands for reforms. This became a compelling strategy that would be adopted by republicans as well. Indeed, two of the most significant events that would galvanize international scrutiny and insistence for regional reforms were the 1972 Bloody Sunday attacks on civil rights protesters and the IRA/INLA's 1981 Hunger Strikes.

Bloody Sunday, 1972

On Sunday, January 30, 1972, a British Parachute Regiment attacked a bipartisan civil rights march in Londonderry/Derry, killing 13 unarmed participants and injuring another 13.[11] Memorialized in music and film (e.g., Irish rock band U2's song "Sunday Bloody Sunday" and Paul Greengrass' film *Bloody Sunday*), the assault in the Bogside neighborhood ignited turmoil throughout the region (see Maps 4.1 and 4.2).

Immediately following the attacks, there was a significant increase in republican paramilitary enlistment and an outpouring of support for Sinn Féin (SF) in Northern Ireland (Moloney 2002). As Gerry Adams, the president of northern SF during the Troubles exclaimed, "after Bloody Sunday, money, guns and recruits flooded into the IRA" (2017, 84). The IRA also launched a widespread "retaliation bombing campaign" contributing to further bloodshed throughout the region.

The traumatic events of Bloody Sunday marked a decisive moment for many within the Catholics/nationalist community who turned away from the state for protection from the RUC or loyalist paramilitary attacks. It also galvanized a shared sense of ethnonational identity among the Catholic/nationalist/republican community. As Sara Cobb, the Director of the Center for the Study of Narrative and Conflict Resolution at George Mason University, explains:

> Communal identities in Northern Ireland were further bolstered by the violence and sectarian nature of the Troubles, which despite obvious presence of individuality and fluidity, group identity was "compressed" into a singular identity.
> *(2003, 298)*

11 The fourteenth subsequently died in the hospital.

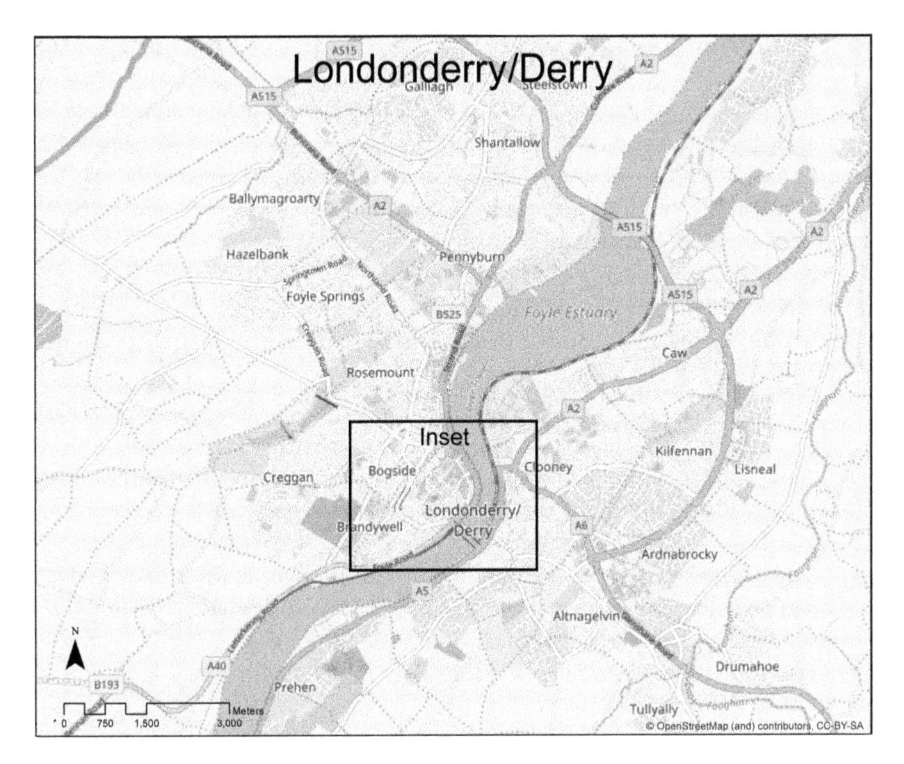

MAP 4.1 Map of Londonderry/Derry

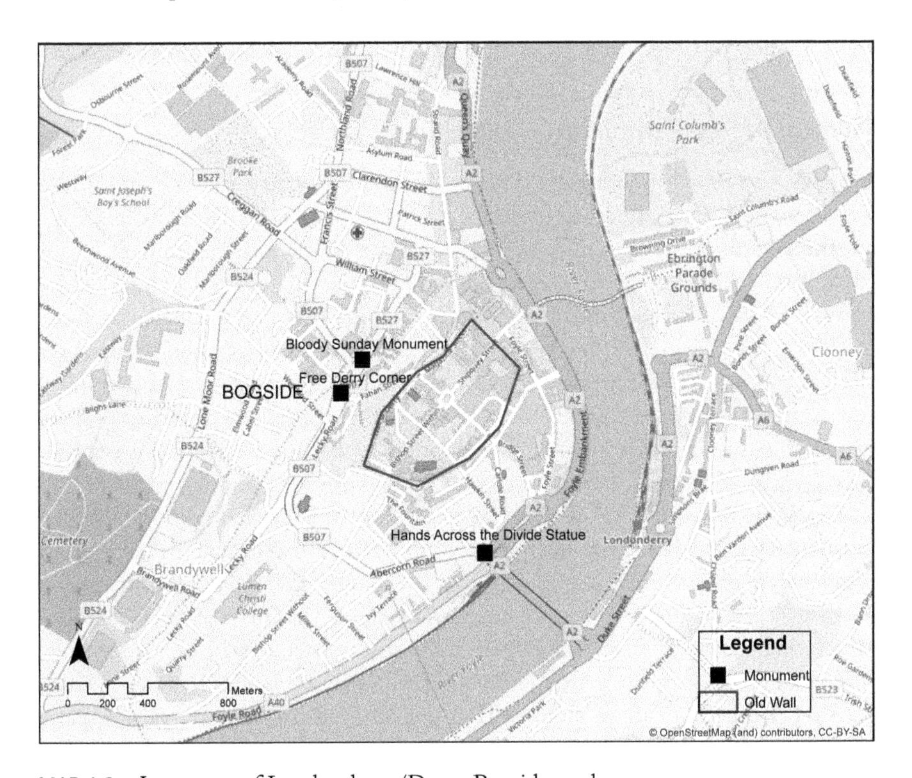

MAP 4.2 Inset map of Londonderry/Derry Bogside enclave

The state exacerbated growing unrest after Bloody Sunday when, after failing to interview all key witnesses, it declared the soldiers were not at fault (Widgery 1972, Conclusion 10). Nationalists/republicans and sympathizers immediately rejected the validity of the inquiry, particularly as all march participants were unarmed. As eyewitness Catholic Bishop Edward Daly explained:

> What really made Bloody Sunday so obscene was the fact that people afterwards, at the highest level of British justice, justified it. I think that's the real obscenity.
>
> *(January 22, 1992)*

The 1972 ruling would eventually be overturned in 2010 when British PM Tony Blair appointed a new inquiry into Bloody Sunday that reported:

> The immediate responsibility for the deaths and injuries on Bloody Sunday lies with those members of the British Support Company whose unjustifiable firing was the cause of those deaths and injuries.
>
> *(House of Commons, Saville Report 2010, Vol. I, Chap. 4, Para. 4.1)*

However, in 1972, the ensuing political strife and region-wide violence fed off of international condemnation of the Bloody Sunday events which placed increasing pressure on Britain to respond. Finally, British PM Edward Heath dissolved the parliament in Stormont in March 1972 and imposed "Direct Rule" in Northern Ireland.

Political protest campaigns – 1981 hunger strike

In addition to Bloody Sunday, the 1981 republican paramilitary prison-hunger strike was a geopolitically significant event during the Troubles. It was part of a multi-decade protest against the criminalization of republican efforts and "Internment." Prior to the 1981 strike, IRA and INLA republican prisoners orchestrated a hunger strike in 1972 to raise awareness of their efforts to pressure the state to categorize them as *political prisoners* instead of *criminals*. Official identification as political prisoners would legitimize republican claims of fighting a political, not criminal, anti-colonial war against the UK. As a result of an IRA ceasefire, the British Secretary of State granted the prisoners their desired "special category status" (i.e., political prisoners), thereby removing the criminal component of their prison categorization.

However, this status was revoked in 1976, thus recriminalizing the republican paramilitary campaign. Prisoners protested this categorical change by refusing to wear prison uniforms. Without clothing, they were often covered with prison blankets, earning them the nickname "blanketmen" in the "Blanket Protests." When their demands to reinstate the special category status went unheeded, they refused to bathe or utilize prison latrines in the "Dirty Protest" (Hennessey 2014).

After five years of these protests, IRA and INLA republican inmates began a hunger strike in March 1981 in the Long Kesh Detention Centre (later renamed

Her Majesty's Prison Maze). The strike ended in October after the deaths of 10 republican strikers. Since the participants were detained in the prison's "H-block" shaped hallways, the letter "H" became a symbol of support for imprisoned paramilitaries.

The strikers staggered their starting dates to extend the duration of their protest and increase political pressure on the government. Simultaneously, republicans outside of the prison organized marches to publicize the strike, generate sympathy, and compel British PM Margaret Thatcher (1979–1990) to grant the prisoners special category status. Their propaganda campaign featured IRA inmate Bobby Sands (imprisoned for gun possession – but not murder) as the "public face" of the protest. Promotional photographs depicted Sands with a large smile, long hair, and often accompanied by lines of his poetry. Indeed, "the fact that he looked more like a drummer in a rock band than a ruthless terrorist was important in the propaganda battle" for the IRA/INLA (McKittrick and McVea 2002, 143). Additionally, reports from visitors to the H-block garnered public sympathy. For example, when the Archbishop of Armagh met with H-block prisoners, he described inhuman cell conditions and inmates' reports of widespread verbal, physical and physiological abuse (O'Fiach 1978).

Significantly, since these protests transpired in a pre-9/11 geopolitical environment, there was more public support for framing the IRA/INLA as an anti-colonial force than what might be possible in a post-9/11 (i.e., post-international) world (Gallaher 2012). Prior to 9/11, many paramilitary organizations narrated their violent actions as anti-colonial actions against oppression and tyranny. In this geopolitical climate, the IRA/INLA's strategically constructed propaganda campaign garnered widespread sympathy, particularly throughout Catholic communities in Northern Ireland.

Despite international pressure from organizations such as the European Commission of Human Rights, Thatcher refused to concede the strikers' demands. Her response affirmed her steadfast position:

> We are not prepared to consider special category status for certain groups of people serving sentences for crime. Crime is crime is crime, it is not political.
>
> *(cited in BBC 2006)*

However, state archival documents reveal that the government had established secret "back channel" negotiations with the IRA to discuss the hunger strike. In contrast to her public statement:

> Mrs. Thatcher's hand was literally all over the "deal" sent to the Provisionals [IRA] – revealing the key involvement of a Prime Minister who claimed that she refused to negotiate with terrorists.
>
> *(Hennessey 2014, 8)*

As the strike continued, its prominence and political influence intensified within the region and beyond Northern Ireland, particularly when Sands became a British Member of Parliament (MP). Amid the hunger strike, an MP for Fermanagh and South Tyrone died unexpectedly, vacating a seat in the British Parliament. Capitalizing on the opportunity, republicans ran Sands as a member of the "Anti H-Block/Armagh Political Prisoner" party for the open seat. While incarcerated by the state, Sands won the election and became an MP of the British House of Commons after he slipped into a starvation-induced coma. He died soon after, ending his 66-day hunger strike (Hennessey 2014). Perceived by many of his supporters as a martyr, Sands' death galvanized widespread international support for the protest, spurred riots, and escalated violence throughout Northern Ireland.

Steve McQueen's film, *Hunger* (2008), depicts much of this turmoil through its examination of the treatment of H-block prisoners and events surrounding their hunger strike. This gritty portrayal focuses on Sands and his death's subsequent impact. During an exchange between Sands and a visiting Catholic priest, Sands explains how his eventual death will be a socio-political mechanism to force change in Northern Ireland:

Don (priest): What is your statement by dying? Just highlighting British intransigence?

Sands: My life means everything to me. Freedom means everything . . . I believe that a united Ireland is right and just . . . a desire for freedom, and an unyielding love for that belief means I can see past any doubts I may have. Putting my life on the line is not just the only thing I can do, Don. It's the right thing. Time's come . . . International pressure on the Brits and all that. Even the Pope's having a say, getting involved. The whole world is trying to get Maggie Thatcher to back down, give us our demands. We're starting a new hunger strike on the first of March. Get the word out . . . If it comes to our dying, I guarantee there'll be a new generation of men and women, even more resilient, more determined.

Significance of the hunger strike

As the film's depiction suggests, MP Sands' death generated international condemnation of the British state's "inflexibility" for the strike. According to Gerry Adams, the hunger strike "had a greater international impact than any other event in Ireland in my lifetime" (2017, 274). While the IRA/INLA prisoners did not gain special status, the strike provided republicans a means to gain media attention and electoral support. This, in turn, brought SF to the political forefront in Northern Ireland as well. Indeed, after the strike, SF ended its refusal to enter mainstream British politics (as an interdiction of British rule) and took their elected seats in Westminster.

Sinn Féin capitalized on its newfound electoral support and launched a new strategy of "Armalite (i.e., arms) and the ballot box," which many consider the "genesis of what would eventually become the peace process" (McKittrick and McVea 2002, 143). More specifically, by finally agreeing to politically engage with the state, SF began a process of modification that ultimately brought the party "out of the shadows" and into mainstream politics. As SF began this transition, John Hume (SDLP) and Gerry Adams (SF) met in secret over several years to discuss peace. Simultaneously, because Thatcher refused to engage with SF publicly, the state initiated behind-the-scenes peace talks with republicans that laid the groundwork for eventual peace within Northern Ireland. Indeed, including republicans in peace negotiations proved fundamental to its success (Purdy 2013).

Adams, who had initially supported the "Long War," eventually helped convince the IRA of the importance of engaging in politics instead of solely relying on violence. Once the IRA accepted the impossibility of winning the "Long War," SF helped negotiate the 1994 paramilitary ceasefire and join talks that eventually led to the 1998 peace agreement. Notably, earlier peace efforts only included the UUP and SDLP and were derailed by paramilitary violence, unionist/loyalist fears of the Republic's involvement in Northern Ireland, and disagreements over paramilitary decommissioning[12] and power-sharing governance. Bringing SF into peace efforts after the 1981 hunger strikes was essential for future peace as it eventually encouraged republican paramilitaries to join negotiations (Kee 2003). Ultimately, UUP leader David Trimble and SDLP leader John Hume jointly received the Nobel Peace Prize in 1998 for their cooperative work for peace in Northern Ireland.

State scale – the United Kingdom

While violence during the Troubles was concentrated in Northern Ireland, the role of the state, how it framed the conflict, and the larger geopolitical context through which Britain perceived the tumultuous years are significant. During the Troubles, Northern Ireland was "the most dangerous place in the world for a British soldier" (Hennessey 2014, 2). Operation Banner stationed over 27,000 British military personnel in Northern Ireland in what became the army's longest continuous campaign.

Despite this, little information about the Troubles or the socio-political environment that fueled it was available throughout the UK outside of Northern Ireland (e.g., O'Hagan 2018). Some of the informational dearth was due to state's media ban on images and interviews with paramilitaries. Additionally, because many perceived SF as the political wing of the IRA, Thatcher barred all photos and public statements from the republican party.

12 Unionists demanded the decommission of paramilitary arms before initiating negotiations, hampering SF's willingness to join peace talks.

The state-funded British media also overlooked many of NICRA's civil rights campaigns or the socio-political discrimination that motivated their protests. Early reporting on the Troubles referred to Northern Ireland as "the province." The allusion to a remote outpost of the Roman empire revealed how many perceived Northern Ireland as divergent and disparate from the rest of the UK. Arguably:

> the civil rights era of the late 60s and early 70s has been conveniently overlooked, because it illuminates the fact that the endemic discrimination in Northern Ireland was ignored for decades by successive Westminster governments of every political hue.
>
> *(O'Hagan 2018)*

In essence, most British living outside of Northern Ireland believed the Troubles resulted from the IRA's archaic and religious struggle to reunite Ireland against the more modern Britain. Appalled, or even apathetic, most interpreted the turmoil as an affront to British self-perception of its democratic and mainly secular culture.

Britain during "Direct Rule"

When the state finally decided Stormont and the RUC could no longer control the escalating violence in Northern Ireland, its "Direct Rule" policy in 1972 brought Westminster to the forefront of the power struggle over Northern Ireland. The EU pressured Britain, a member state from 1973 to 2020, to resolve the conflict, particularly as media reports (e.g., covering Bloody Friday, Enniskillen Bombing, Bloody Sunday, the 1981 Hunger Strike) fomented international outrage. However, its intervention occurred during a tumultuous time for the state. Ensnared in the Cold War (including the Vietnam Conflict) and internal anti-war protests, paranoia over Soviet espionage, the 1973 Arab-Israeli war, and the OPEC oil embargo, its efforts to balance competing political forces within Northern Ireland faltered during various significant geopolitical moments including the Sunningdale Agreement, Falklands War, and the Anglo-Irish Agreement.

Sunningdale

In an effort to restore self-government in the region, the state forged the Sunningdale Agreement in December 1973. This agreement proposed a power-sharing, devolved administration (the Northern Irish Executive) that included an "Irish dimension." This granted the Republic of Ireland an advisory role in internal matters in Northern Ireland through the creation of the Council of Ireland. Predictably, unionist/loyalist communities opposed the agreement based on the Republic's involvement in Northern Irish affairs.

During the unionist/loyalist protests, the state was distracted by the final years of the Vietnam conflict and negotiating the UK Miners' Strike (1984–1985). This strike, the longest industrial dispute in British history, transpired when the state imposed pay caps that aggravated labor unions. In response, the unions reduced coal production to limit available energy throughout the state. British PM Edward Health subsequently enforced a "Three Day Work Order" restricting commercial energy consumption to three consecutive workdays a week. With this state-wide crisis at the forefront of his political agenda, the preoccupied PM did little to support Northern Ireland's Chief Executive, Brian Faulkner's efforts to regain control of Northern Ireland during the Sunningdale protests. Ultimately, the loyalist-organized anti-Sunningdale Ulster Workers Council strike paralyzed the region and Faulkner resigned in 1974. The Sunningdale government collapsed the following year.

The state was also unsuccessful in its early attempts to negotiate peace with paramilitaries. For example, during the "Long War" some of the IRA's leadership secretly reached out to the state for potential peace talks. The state initially declined to negotiate due to mistrust. Tensions increased during Thatcher's appointment as PM (1979–1990) as her Northern Irish policies were designed to destroy republicanism instead of seeking a peaceful solution. Her unrelenting policy during the 1981 Hunger Strike further exacerbated Irish republicans and their international sympathizers.

Falklands War

The outbreak of the Falklands War in 1982 further exacerbated turmoil. For example, some of the British troops stationed in Northern Ireland were reassigned to the Falklands. Their departure placed additional stress on the remaining security forces. Additionally, political relations between the UK and the Republic of Ireland soured when British troops sank an Argentine ship and the Irish government described the attack as an assault by British "aggressors." The Republic then called a UN Security Council meeting to discuss appropriate censure of the UK (Borders 1982).

The Republic's geopolitical posturing over the Argentine ship outraged Thatcher. The PM interpreted it as an indication of the Republic's acceptance of Argentine control of the Falklands. This strained relations between the two states for years and hindered a collaborative solution for Northern Ireland. Indeed, the Republic's UN ambassador described it as the "greatest single controversy in Anglo-Irish relations for a decade" (O'Connell 2012).

Anglo-Irish Agreement

Despite numerous impediments, peace negotiations for the conflict in Northern Ireland continued. In 1985, British and Irish governments signed the Anglo-Irish Agreement (AIA), laying the foundation for future peace in the region. The AIA assured that the potential dissolution of Northern Ireland's constitutional link to the UK would only occur with the consent[13] ("consent principle") of the majority

13 This clause was also previously included in the Northern Ireland Constitution Act of 1973 and the failed Sunningdale Agreement.

of the region's inhabitants. It also granted the Republic a significant "consultative role" in the legislation and management of Northern Irish affairs.

While moderate parties in Northern Ireland supported the AIA, the political extremes staunchly opposed it. For example, loyalists refused to accept a power-sharing government or the Republic's involvement in Northern Irish matters. The UUP and DUP orchestrated massive protests against an agreement that they believed weakened Northern Ireland's union within the UK. DUP leader, Ian Paisley, lashed out against the AIA, which he perceived as Thatcher's betrayal of Northern Ireland as it permitted the Republic an advisory role regarding the governance of the region.

Simultaneously, most republicans opposed the AIA because it preserved Northern Ireland within the UK. They also feared that new cooperation between the states would target the IRA in Northern Ireland. However, some SF leaders interpreted the AIA as a pivot away from Britain's sole loyalty to the unionists, and this political nod encouraged the party to begin to modernize its strategy for the region (McKittrick and McVea 2002). Despite the controversies, the intra-governmental collaborations forged through the AIA became the foundation of a peace process culminating in the 1998 peace agreement for Northern Ireland.

International scale – state and supranational players

While many foreign states affected events during the Troubles, the following section briefly describes some of the most significant external geopolitical actors: The Republic of Ireland, Libya, the United States, and the supranational European Union.

The Republic of Ireland

The Irish state's involvement in Northern Irish matters increased during the Troubles. After the collapse of the Sunningdale Agreement's power-sharing executive and the Council of Ireland, the state continued to work with Westminster for a resolution in Northern Ireland. In May 1980, PM Thatcher and Irish Taoiseach Charles Haughey (1979–1981; 1982; 1987–1992) met in Dublin. Commonly referred to as the "teapot summit," a reference to Haughey's gift for Thatcher, it was the first meeting of heads of these states since the Anglo-Irish Treaty in 1921. While Thatcher perceived it as a mere meeting, Haughey publicly described it as the beginning of a political dialogue regarding a constitutional change in the status of Northern Ireland.

When Garrett FitzGerald of Fine Gael replaced the staunchly republican Haughey of Fianna Fail in 1981 as Taoiseach, he imparted an increasing willingness to collaborate with Britain for a peaceful resolution to the Troubles. His approach provided the foundation for negotiations that produced the AIA. FitzGerald was not only motivated by the desire for peace in Ireland but also feared a protracted conflict would foster the spread of northern SF's radicalism into the Republic and eventually destabilize the Irish state (FitzGerald 1991).

Indeed, by the 1980s, many within the Republic disapproved of the IRA in Northern Ireland. For example, in the early years of the Troubles, after violent attacks on "Catholic" neighborhoods in Northern Ireland, some of the Republic's politicians and public supported the idea of providing funding and arms to the northern IRA. In 1969, the state established refugee centers throughout the Republic's borderlands for Irish fleeing Northern Ireland. As Taoiseach Lynch (1966–1973; 1977–1979) declared, action was necessary as the "Irish government can no longer stand by" (Lynch, CAIN recording 1969).

However, widespread support for the IRA in the Republic dissolved after it attacked a British embassy in Dublin in 1972. This violent and geopolitically motivated attack transformed much of the Irish public's perception of the IRA, no longer seeing the paramilitaries as anti-colonial freedom fighters, instead of considering the IRA a terrorist organization (Patterson 2013). The Irish public's abhorrence of IRA violence intensified after the bombing of a WWI commemoration that killed 11 civilians in Enniskillen, Northern Ireland.

Enniskillen Bombing

The 1987 Enniskillen Bombing ("Poppy Massacre") is considered a key turning point in the Troubles. Prior to this bombing, many republicans interpreted Protestant/unionist commemorations of WWI as symbolic of British oppression and its colonial past. However, the IRA's attack on civilians honoring those who died in the war outraged the Irish public and compelled the Irish state to re-evaluate how it acknowledged its cooperation with Britain during WWI (Robinson 2010).

In Dublin, over 50,000 members of the public waited for hours to sign a book of condolence sponsored by the mayor. Sympathizers filled 40 books with words of sorrow and support for the victims of the Enniskillen attack (Mallie and McKittrick 1996, 60). The Catholic Church condemned the IRA and preached it was a sin to support the organization (Cooney 1987). Irish elected officials unanimously expressed indignation for the bombing (House of Commons 1987). Most notably, in regard to supranational cooperation in the peace process, Taoiseach Haughey demanded the culprits be "utterly repudiated and brought to justice." He also wrote PM Thatcher to suggest "combining all security forces on this island for an all-out effort to have the perpetrators brought to justice" (CJ 4/6872 1987).

Subsequent cooperation between Irish and British governments also focused on peace talks and resulted in the 1993 Downing Street Declaration[14] (DSD). As part of this agreement, the Republic accepted the "consent principle" of Northern

14 The DUP condemned the DSD (and abstained from signing the 1998 peace agreement). According to Ian Paisley of the DUP, the Republic's involvement in Northern Irish politics and the British government's efforts to include SF in discussions was the "worst crisis in Ulster's history" (Paisley 1994).

Irish citizens' right to decide the political future of the region. This was significant because DSD was intended to:

> remove the causes of conflict, to overcome the legacy of history, and to heal the divisions which have resulted – recognizing that the absence of a lasting and satisfactory settlement of relationship between the people of both islands has contributed to continuing tragedy and suffering.
>
> *(CAIN 1993)*

Thus, this geopolitical capitulation was the Republic's gesture of appeasement to unionist/loyalists' concerns about its previous constitutional claim of geopolitical governance of Northern Ireland.

Libya

During the Troubles, the IRA strengthened its armaments with the support of Libyan dictator Muammar Gaddafi. Throughout the 1970s, Gaddafi supplied the IRA with weapons (including anti-aircraft guns and SAM-7 missiles) and financial support due to Libya's deteriorating relations with the UK. Subsequently, the IRA became one of the most-armed paramilitary organizations in the world at the time. With these supplies, the IRA targeted British troops, police, and "spectaculars" (high profile targets) throughout the UK and mainland Europe.

However, Libyan support for the IRA rapidly dwindled after the Enniskillen Bombing. As a result of the civilian deaths in the attack (instead of British politicians or military troops), the Libyan government condemned the IRA and terminated its relationship with the IRA. The Libyan state's searing declaration: "Libya is aware of the difference between legitimate revolutionary action and terrorism aimed at civilians and innocent people. This action does not belong to the legitimate revolutionary operation" prompted the IRA to publicly apologize "with deep regret" for the attack.

Not only did the Enniskillen Bombing hurt the IRA's claim that its actions were "anti-colonial" but it also severed the IRA's Libyan arms and ammunition pipeline that was necessary to maintain its operations. Libya's dissolvement of its previous relationship with the IRA may represent the most significant blow to the IRA's "Long War" against the British state (Moloney 2002). Incidentally, the loss of Libyan support (politically and militarily) for their violent actions encouraged many republicans to consider working toward a peaceful resolution. Indeed, after Gerry Adams (SF) publicly condemned the Enniskillen Bombing, he agreed to resume peace talks with John Hume's SDLP. Hume and Adams began meeting in secret to discuss "another alternative way forward" (Black and Young 2020). As previously stated, their covert cooperation spanned a decade and contributed to the multiscalar networked foundation that would ultimately produce the Irish peace process in the 1990s (Jackson 2003).

The United States (US)

Under the Clinton administration (1993–2001), the US played a key role in Northern Ireland's peace process. Previously (particularly during Reagan's presidency), the US did not engage in inter-British politics. However, support for US involvement in Northern Ireland grew as the American public learned about the Troubles through international media coverage that highlighted parallels between America's and NICRA's civil rights movements, attacks on NICRA marches, and republican prison strikes. In 1971, American Senator Ted Kennedy gave a speech arguing that "Ulster is becoming Britain's Vietnam" and began collaborating with John Hume (SDLP) for a peaceful resolution to the Troubles.

As a result of the mounting pressure, particularly from prominent Irish Americans, President Clinton abolished former presidents Reagan and Bush's laissez-faire approach to the Troubles. The US became an external "outside the box" mediator between the UK and the Republic. Indeed, since Irish and British governments (as well as regional/local stakeholders and paramilitaries) were deadlocked in negotiations, the US served as an apparently impartial actor that brought inimical forces together for peace proceedings.

Despite British opposition, in 1994 Clinton invited Gerry Adams to speak with him in Washington about the Troubles. Outside of the UK, Adams circumvented the British media ban on republicanism and spoke openly about the treatment of Catholics within Northern Ireland. In November 1995, Clinton appointed Senator George Mitchell to chair a committee on the peace process in Northern Ireland. The committee encouraged immediate initialization of multi-party peace talks prior to paramilitary decommissioning, a significant concession for unionists/loyalists.[15] Despite major setbacks, including the collapse of the 1994 paramilitary ceasefire and the IRA bombing of London's Canary Wharf in 1996, Mitchell collaborated with the SDLP, Alliance Party, and unionists to forge ahead with negotiations. With the added support of the new British PM Tony Blair (1997–2007), who supported the peace process, representatives engaged in peace talks that resulted in the 1998 peace agreement.

Supranational scale – the European Union (EU)

The EU played a key role in facilitating the 1998 peace agreement. As both the Republic of Ireland and the UK were EU member states prior to Brexit, the supranational organization's protocols restructured political and economic governance policies between the two states. For example, the creation of the European Single Market Policy in 1992 stimulated the movement of goods, services, and economic capital within the economic union, including across the Irish border. However,

15 British PM Tony Blair would finally sideline the requirement of prior arms decommissioning in 1997 in order to push for an eventual peace agreement.

notable economic disparity between the two political units in Ireland remains. This is primarily a reflection of the UK's "monetary and fiscal sovereignty" when it declined to join the Eurozone (Patterson 2013).

Despite these distinctions, EU integration was a significant driving force for several cross-border initiatives between Northern Ireland and the Republic. For example, European integration underpinned the transformation of the Irish border from militarized to inconspicuous. Despite the UK not joining the EU's Schengen "common border" area, which should therefore require a "hard" border in Ireland, the 1998 peace agreement forged a "softer" border, fostered by the spirit of EU integration.

The EU also financially supported peacebuilding organizations and legislative change in Northern Ireland with the Special Support Program for Peace and Reconciliation or Peace I Fund (1995–1999), Peace II (1999–2006), and Peace III (2007–2015). These grants funded reconciliation and cross-community programs and agencies. It also worked to address key structural and psychological sources of conflict in the region. In 2014, the EU launched Peace IV (2014–2020) specifically to support children in post-conflict Northern Ireland and the border region in Ireland. This most recent program includes funding for cross-communal education as well as shared spaces and services. Chapter 7 provides an in-depth analysis of a shared community center in a sectarian part of Belfast that has benefitted from the EU Peace initiative. Thus, the EU reinforced the political machinery underpinning cooperation and reconciliation at the supranational, national, regional, and local scales.

The peace agreement, a multiscalar effort, and the fragile road forward

Ultimately, the US chaired the negotiations between the Northern Irish political parties, the British state, and the Irish government. These negotiations were built on foundations forged by the AIA in 1985 and subsequent negotiations, including the secret "back channel" discussions between the British state and republicans. When the peace proposal was first presented to the participating regional parties (the DUP left the peace talks when SF was included), unionist parties rejected it. Some of the more controversial elements of the agreement included paramilitary prisoner releases – a strategy designed to encourage paramilitaries to support the agreement and ultimately proved successful. Unionists' initial rejection of the proposal prompted PM Blair and the Irish Taoiseach Bertie Ahern (1997–2008) to travel together to Stormont to encourage dissenters to accept the agreement. Simultaneously, President Clinton spoke with regional party leaders to convince them to support the agreement as well.

All participating parties accepted the peace agreement the following day. The agreement was ratified in Northern Ireland in May 1998 (71% support by the electorate) and in the Republic of Ireland (supported by 94% of voters). It facilitated the return of regional governance through the creation of a new devolved power-sharing parliament, the Northern Irish Assembly. Government mandates safeguarded cross-community cooperative governance by requiring that no one party can have sole control of the assembly. This limitation encourages political collaboration through the basis of parallel consent (i.e., support from majority of both nationalists and unionists).[16]

The ramifications of the agreement extend beyond regional borders to include, for example, alterations of the Irish Constitution. The Republic of Ireland, which historically insisted on the necessity of a united Ireland under "Irish" rule, agreed to modify its constitution to remove its territorial claim to the six counties that constitute Northern Ireland. Both British law and the Republic of Ireland's constitution now recognize the "principle of consent" that assures the political future of Northern Ireland will be determined by its citizens through a regional referendum. These governments also agree to allow Northern Irish citizens the right to hold British or Irish passports – or both. The agreement established a political body that connects the UK's "devolved assemblies" with the Republic and other cross-border cooperative institutions that bridge the two states' governments.

Despite these successes, the political road forward for Northern Ireland is a fragile one. For example, the 1998 peace agreement did not attempt to foster a newly unified nation within Northern Ireland. Instead, the text recognizes the existence of two distinct ethnonational communities, thereby reinforcing binary perceptions of identity in Northern Ireland (e.g., Wilford and Wilson 2006). Chapter 7 discusses how many cross-community advocates believe this geopolitical reinforcement of social bordering continues to harm reconciliation efforts in Northern Ireland.

Despite its recent reinstatement, the devolved legislative assembly in Northern has been suspended intermittently since 1998. In 2006, the 1998 peace agreement was amended by the St. Andrews Agreement. The new legislation mandates that the regional power-sharing government include the two competing political extremes, the DUP and SF. This modification reflects the growth in popularity of the more radical DUP, which became more prominent than the more moderate UUP, just as republican SF has surpassed moderate SDLP's political prominence in Northern Ireland. This may suggest a growing polarization within the Northern Irish community as contestation over segregation, political marches, policing, sectarian-divided education, and employment equity remains.

Dissent paramilitary violence also continues to haunt Northern Ireland, e.g., Omagh bombing in 1998 by the Real IRA (RIRA), a splinter group who opposed the peace agreement). Small splinter IRA groups' acts of violence include the "New" IRA's April 2019 murder of Northern Irish journalist Lyra McKee in Derry/Londonderry (O'Loughlin 2019). Indeed, the legacy of violence, division,

16 For a detailed description of the agreement and parliament functions, see McKittrick and McVea 2002.

and hatred maintains a stronghold for many in Northern Ireland, even after the peace agreements. As Queen's University professors in Belfast argue:

> by 1998, thirty years after the conflict started, one in seven of the population reported being a victim of violence; one in five had a family member killed or injured; and one in four had been caught up in an explosion.
>
> *(Hayes and McAllister 2001, 901)*

This is particularly true in certain geographically concentrated communities where sectarian violence intermittently erupts within the region (e.g., from 1999 to 2014, there were over 2,000 shootings and 1,600 bombings) as Northern Ireland continues to evolve in its delicate and multiscalar transition to peace (Gaffikin et al. 2016).

Examination of Northern Ireland's ethnonational conflict across various scales fosters a nuanced understanding of the dynamic spatial contexts through which the Troubles and peace process evolved. Contestation and cooperation develop at various scales and stretch across interwoven spaces of discrimination, violence, and peace work. Despite the peace agreement, enduring divisions and competing territorial claims remain in certain sectarian enclaves, such as northwest Belfast. The following two chapters provide insight into how patterns of segregation and division play out spatially on the local landscape in this contested part of the capital city.

References

Adams, Gerry. 2017. *Before the Dawn: An Autobiography*. Dublin: Brandon.

Barrett, S. 2013. "The Necessity of a Multiscalar Analysis of Climate Justice." *Progress in Human Geography* 37 (2): 215–33.

Barry, John. 2003. "National Identities, Historical Narratives and Patron States in Northern Ireland'." In *Political Loyalty and the Nation-State*, 189–205. Oxfordshire: Taylor and Francis.

BBC. 2006. "What happened in the Hunger Strike?" http://news.bbc.co.uk/2/hi/uk_news/northern_ireland/4941866.stm

Black, Rebecca, and David Young. 2020. "Gerry Adams Pays Tribute to John Hume as a 'Giant in Irish Politics.'" www.belfasttelegraph.co.uk/news/northern-ireland/gerry-adams-pays-tribute-to-john-hume-as-a-giant-in-irish-politics-39419279.html.

Borders, William. 1982. "Falkland Crisis Is Staining Irish-British Relations." *The New York Times*.

BSSRS (British Society for Social Responsibility in Science). 1974. *The New Technology of Repression: Lessons from Ireland*. London: BSSRS.

CAIN. 1993. "Joint Declaration on Peace: The Downing Street Declaration." December. https://cain.ulster.ac.uk/events/peace/docs/dsd151293.htm.

Cairns, Ed, and John Darby. 1998. "The Conflict in Northern Ireland: Causes, Consequences, and Controls." *American Psychologist* 53 (7): 754–60.

Cobb, Sara. 2003. "Fostering Coexistence in Identity-Based Conflicts: Towards a Narrative Approach." *Imagine Coexistence: Restoring Humanity after Violent Ethnic Conflict*: 294–310.

Cooney, J. 1987. "Ireland's Catholic Bishops Condemn IRA." *The Times*, November 14, 1987.

Dempsey, Kara E. 2016. "Competing Claims and Nationalist Narratives: A City/State Debate in a Globalising World." *Tijdschrift Voor Economische En Sociale Geografie* 107 (1): 33–47.

Devlin, Bernadette. 1969. *The Price of My Soul*. New York, NY: Knopf.

Farrell, Michael. 1980. *Northern Ireland: The Orange State*. London: Pluto Press Limited.

FitzGerald, Garret. 1991. *All in a Life: Garret FitzGerald, an Autobiography*. Dublin: Gill & Macmillan.

Flint, Colin. 2016. *Geopolitical Constructs: The Mulberry Harbours, World War Two, and the Making of a Militarized Transatlantic*. Lanham, MD: Rowman & Littlefield.

Gaffikin, Frank, Chris Karelse, Mike Morrissey, Clare Mulholland, and Ken Sterrett. 2016. *Making Space for Each Other: Civic Place-Making in a Divided Society*. Belfast: Queen's University Belfast.

Gallaher, Carolyn. 2012. "Terrorism." In *Key Concepts in Political Geography*, edited by Carolyn Gallaher, T. Dahlman, Mary Gilmartin, Alison Mountz, and Peter Shirlow, 247–59. London: Sage.

Gregory, Ian N., Niall A. Cunningham, Paul S. Ell, Christopher D. Lloyd, and Ian G. Shuttleworth. 2013. *Troubled Geographies: A Spatial History of Religion and Society in Ireland*. Bloomington: Indiana University Press.

Hayes, Bernadette, and Ian McAllister. 2001. "Sowing Dragon's Teeth: Public Support for Political Violence and Paramilitarism in Northern Ireland." *Political Studies* 49: 901–22.

Hennessey, Thomas. 2014. *Hunger Strike: Margaret Thatcher's Battle with the IRA, 1980–1981*. Kildare: Irish Academic Press.

House of Commons. 1987. "Debate over security question in Northern Ireland as discussed by the House of Commons at Belfast, Northern Ireland, on May 6, 1987." Debate. C735. https://www.theyworkforyou.com/debates/?id=1987-05-06a.735.0

________. 2010. *Saville Report, Official Bloody Sunday Inquiry*. London: The Stationery Office.

Jackson, Alvin. 2003. *Home Rule: An Irish History, 1800–2000*. Oxford: Oxford University Press.

Kee, Robert. 2003. *Ireland: A History*. London: Abacus Books.

Lynch, John. 1969. "A Review of the Situation in the Six Counties of North-East Ireland and a Statement of the Irish Government's Policy as given by the Taoiseach (Prime Minister of Ireland) in a Speech at Tralee, Co. Kerry, on September 20, 1969." Speech. Tralee, Co. Kerry. https://cain.ulster.ac.uk/othelem/docs/lynch/lynch69.htm.

Mallie, E., and David McKittrick. 1996. *The Fight for Peace*. London: Heinemann.

Marston, Sallie A., and Neil Smith. 2001. "States, Scales and Households: Limits to Scale Thinking? A Response to Brenner." *Progress in Human Geography* 25 (4): 615–19. https://doi.org/10.1191/030913201682688968.

McDowell, Sara, and Peter Shirlow. 2011. "Geographies of Conflict and Post-Conflict in Northern Ireland: Conflict in Northern Ireland." *Geography Compass* 5 (9): 700–9. https://doi.org/10.1111/j.1749-8198.2011.00444.x.

McKittrick, David, and David McVea. 2002. *Making Sense of the Troubles: A History of the Northern Ireland Conflict*. Chicago: New Amsterdam Books.

McQueen, Steve. 2008. *Hunger*. Film. Pathe Distribution.

Mesev, Victor, Peter Shirlow, and Joni Downs. 2009. "The Geography of Conflict and Death in Belfast, Northern Ireland." *Annals of the Association of American Geographers* 99 (5): 893–903. https://doi.org/10.1080/00045600903260556.

Moloney, Ed. 2002. *A Secret History of the IRA*. New York, NY: W & Norton and Company.

O'Connell, Hugh. 2012. "Falklands War: Irish Response to Belgrano Sinking Drew British Anger." *The Journal*, May 12, 2012. www.thejournal.ie/anglo-irish-relations-falklands-war-belgrano-723142-Dec2012/.

O'Fiach. 1978. "Archbishop O'Fiach's Statement on Maze Prison." August 1. Report. https://cain.ulster.ac.uk/proni/1978/proni_NIO-12-68_1978-08-01_a.pdf.

O'Hagan, Sean. 2018. "Northern Ireland's Lost Moment: How the Peaceful Protests of '68 Escalated into Years of Bloody Conflict." *The Guardian*, April 22, 2018.

O'Loughlin, Ed. 2019. "New IRA Apologies for Killing Journalist in NI." *The New York Times*, April 23, 2019. www.nytimes.com/2019/04/23/world/europe/lyra-mckee-new-ira-apology.html.

Paasi, Anssi. 2004. "Place and Region: Looking Through the Prism of Scale." *Progress in Human Geography* 28 (4): 536–46.

Patterson, Henry. 2013. *Ireland's Violent Frontier: The Border and Anglo-Irish Relations during the Troubles*. New York, NY: Springer.

Purdy, Martina. 2013. "Peace in the Troubles." In *BBC History of the Troubles*. www.bbc.co.uk/history/topics/troubles_peace.

Robinson, Helen. 2010. "Remembering War in the Midst of Conflict: First World War Commemorations in the Northern Irish Troubles." *Twentieth Century British History* 21 (1): 80–101.

Sales, Rosemary. 1997. *Women Divided: Gender, Religion and Politics in Northern Ireland*. East Sussex: Psychology Press.

Widgery. 1972. "Report of the Tribunal Appointed to Inquire into the Events on Sunday, 30 January 1972, Which Led to Loss of Life in Connection with the Procession in Londonderry on That Day." Report. https://cain.ulster.ac.uk/hmso/widgery.htm.

Wilford, Rick, and Robin Wilson. 2006. *The Trouble with Northern Ireland*. Belfast: Democratic Dialogue.

5

THE WALLS WITHIN

Segregation, peacelines, and flags

Geographic and ethnonational divisions are ubiquitous for many residents of Northern Ireland. For these individuals, one's territorial sense of belonging and identity (i.e., elements of socio-political bordering processes) is often rooted within social networks bound to place. When claims to these places are contested, enmity between the communities can become spatially and violently discernible on the region's landscape. For example, the turmoil during the formation of Northern Ireland triggered a process of ethnonational residential segregation. However, during periods of relative calm (e.g., mid-1930s–mid-1960s), segregation did not decrease; instead, these spatial arrangements became institutionalized for much of the region.

Segregation increased during the Troubles. Intimidation and violence prompted the largest population movement in Europe since WWII as ethnonational communities divided into a fractured patchwork of ethnic enclaves in Northern Ireland (e.g., Mesev, Shirlow, and Downs 2009). The subsequent enclaving is a manifestation of social and spatial bordering practices (i.e., physical and/or perceived boundaries) that reflect and foster continued ethnonational divisions in the region (e.g., Gaffikin et al. 2016). As Irish author, Dervla Murphy writes:

> The average Northern Ireland citizen is born either Orange [Protestant/unionist/loyalist] or Green [Catholic/nationalist/republican]. His whole personality is conditioned by myth and he is bred to live the sort of life that will reinforce and protect the myth for transmission to future generations. Moreover, these myths are used daily to justify distrust and resentment of "the other side."
>
> *(Murphy 1979, 188)*

DOI: 10.4324/9781003141167-5

This chapter examines how ethnonational tensions in Northern Ireland resulted in substantial residential segregation and the construction of urban "peacelines" – walls that divide ethnonational communities – through a local scaled case study of segregated northwest Belfast. This is followed by an investigation of political flagging as a territorial marking practice. This chapter concludes with a reflection of how spatial cleavages and territorial marking foster continued divisions among Northern Irish communities.

Sectarianism and segregation

Spatial analysis demonstrates that after the 1998 peace agreement, over 40% of Northern Ireland's inhabitants continue to live in ethnonationally segregated communities (Nolan 2017). This percentage was higher during periods of sectarian strife, particularly in contested neighborhoods in Belfast, Londonderry/Derry, central-Ulster, and border regions such as south Armagh. These areas also experienced a disproportionately high concentration of violence (Gregory et al. 2013). For example, from 1968 to 1993 over 50% of all Troubles-related killings transpired in segregated enclaved areas of Belfast (Kelters 2013). These neighborhoods were often targeted by armed forces due to their ethnonational affiliations and suffered the highest rates of civilian casualties by gunfire, bombings, riots, and arson.

Research suggests that the increased exposure of sectarian violence raised the enclaved residents' potential for personal bodily harm and cultivated support for paramilitary organizations. More specifically:

> public exposure to violence serves to mould popular attitudes towards the use of violence as a political tool, and in turn to engender further political violence. In other words, violence may not simply be a consequence of other (mainly political) things, but it may feed off itself in a continuous and perpetual cycle.
>
> *(Hayes and McAllister 2013, 901–2)*

Impacted by sectarian violence and segregation, many residents manifested strong territorialized local identities and support for paramilitaries. As paramilitary organizations held a strong presence in these enclaves, fatal shootings by security forces or opposing paramilitaries became increasingly common. Rivalries between "corresponding" paramilitary organizations also erupted periodically, increasing the number of civilian and paramilitary fatalities in these enclaves.

Despite the presence of varying degrees of segregation in Belfast since its founding, residential divisions increased dramatically during the Troubles. By the 1970s, intimidation and violence drove residents into segregated ethnonationally homogenous "sanctuary" enclaves, entrenching residents in enclaves. This was especially true for those who were less economically mobile[1] (e.g., NISRA 2019). Indeed,

1 Neoliberal deindustrialization disproportionately impacted Belfast's working-class neighborhoods, as many were employed in the city's heavy industry or manufacturing and subsequently lost their jobs (Gaffikin and Morrissey 2016).

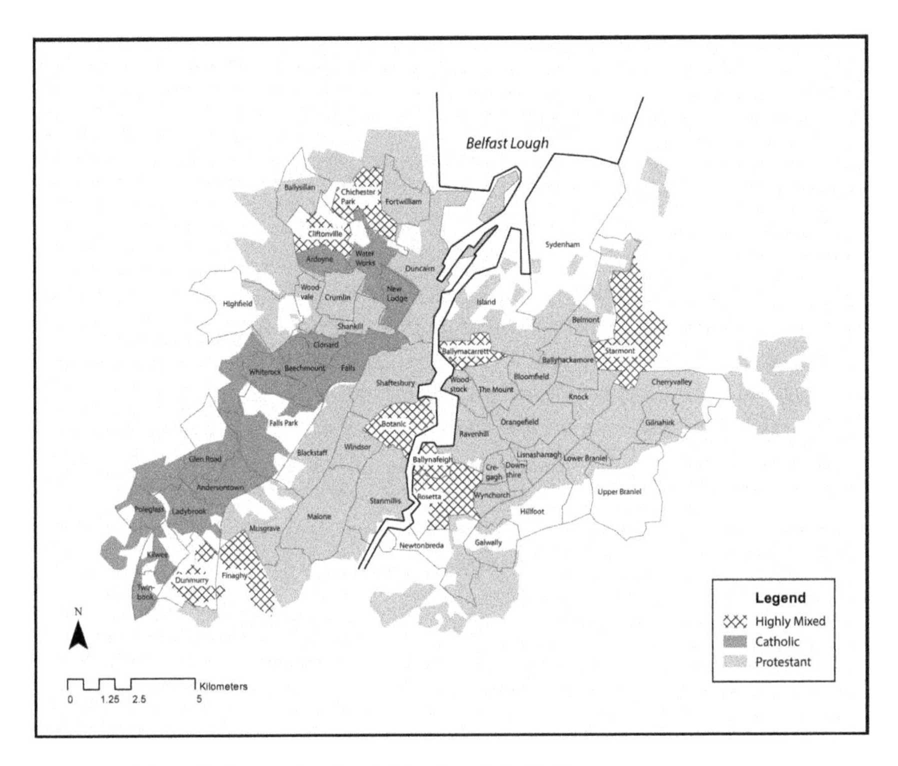

MAP 5.1 Map of ethnonational neighborhoods in Belfast

the most segregated parts of Northern Ireland, such as northwest Belfast – that comprising neighborhoods such as Shankill (Protestant/loyalist, hereafter P), Crumlin (P), and Ardoyne (Catholic/republican, hereafter C), and New Lodge (C) – are consistently the most violent and economically disadvantaged in the UK (Nolan and Bryan 2016).

In this chapter, Catholic/national/republican communities are described using the vernacular categorization – Catholic (C) – and the Protestant/unionist/loyalist communities are identified as Protestant (P). The legacy of the protracted sectarian violence manifests through residential patterns with a high correlation between location and economic opportunities (e.g., Cunningham and Gregory 2014; Nolan 2017) (see Map 5.1).

Film depictions of segregated Belfast: '71

Depictions of these segregated neighborhoods during the Troubles appear in film and television, including the historical-action film, *'71*. The film follows English army recruit, Gary Hook, who is deployed to Belfast to assist the RUC attempt to quell growing unrest in the city. In Belfast, he witnesses widespread turmoil, intimidation, and violence. Catholic mothers in republican enclaves loudly bang trashcan lids on the streets to warn local residents of impending attacks when British forces arrive in armored vehicles. During an aggressive police/military raid, a woman begs the soldiers to leave. She screams, "For God's sake, will you never leave us alone?!," an affirmation of the sense of betrayal and mistrust that many Catholics experienced during the Troubles.

Hook is later separated from his platoon in a volatile sectarian interface area in west Belfast and must try to return to his barracks on foot. While trying to evade capture by hostile paramilitaries, he witnesses ethnonational and inter-sectarian violence, and receives help from both sides of the sectarian divide. He also encounters British spies with ambiguous loyalties, revealing the state's paramilitary proxy wars that complicate simplistic conceptions of Catholics as the "the enemy within." Thus, the film provides insight into some of the complex geopolitical mechanisms that provoked violence and segregation, especially for those living in sectarian ethnonational enclaves.

"Peacelines"

During periods of acute upheaval in the early years of the Troubles, residents and authorities began to physically divide contiguous rival enclaves with partition walls. These walls, commonly referred to as "peacelines" or "peacewalls," were intended to be a provisional buffer from sectarian violence and rioting. As violence or its imminent threat continued, additional peacelines were cursorily erected in other contentious areas. Both ethnonational communities created new peacelines or reinforced previously established ones. Residents believed the barriers would eventually be removed and the army would then forge and maintain its own "Peace Line" along contested areas.

While government agencies[2] removed some of the temporary walls, they also replaced many of the previously makeshift barriers with permanent steel, brick, or concrete peacelines. Some are over 25 feet high and three miles in length. Areas surrounding peacelines are territorially demarcated with ethnonational graffiti and paramilitary insignias. Adjacent streets are often stark as former residents abandoned the area during times of upheaval. Peacelines continue to obstruct mobility and communal interaction, as a material legacy of authorities' acquiescence to sectarian segregation (see Figure 5.1).

Belfast as a case study: walls and segregation

Portions of Belfast's thirteen miles of peacelines that snake across its urban landscape are some of the region's most notorious. While the shortest peacelines are 9.8 feet

2 These agencies include Northern Ireland Office and Belfast City Council.

FIGURE 5.1 Cupar Way peace wall, Belfast, Northern Ireland.

Source: Photo by author.

tall, the highest reach over 40 feet, designed to stop human interaction as well as projectiles hurled throughout the Troubles (e.g., Scarman 1972). During the Troubles, walls were almost exclusively located in lower socioeconomic communities. While Belfast's urban core benefitted from targeted financial investment through urban renewal[3] and tourism, little investment or infrastructural growth arrived in working-class neighborhoods that suffered the highest rate of unemployment (Gaffikin et al. 2016).

Some peacelines are constructed over the foundations of houses previously destroyed by riots and arson. Others extend across streets with barrier gates maintained by police and local community leaders who can close them during turbulent times to block motorized traffic, pedestrians, and projectiles between the communities. These gates are specifically locked during the controversial annual unionist/loyalist "marching season" (April–August), discussed in Chapter 6. In this way, peacelines shape urban transportation patterns and spatially illustrate the legacy of violence during the Troubles.

The Cupar Way wall

Belfast's first and largest wall divides the Lower Falls and Clonard (C) enclaves from Shankill (P) in northwest Belfast, an area that remains one of the most segregated in the city. At its highest point, it is over 40 feet tall and almost a mile long. After its initial construction, additional barriers' sections appeared throughout this contested area of northwest Belfast. In dangerous areas along the Cupar Way peaceline, residents added additional netting, wire fencing, or caging to stop projectiles such as glass bottles, rocks, or Molotov cocktails/petrol bombs (see Figure 5.2). This wall bisects Cupar Way to separate contiguous loyalist Shankill (considered 99% Protestant) and republican Falls/Clonard (considered 98% Catholic) neighborhoods.

The army constructed the first iteration of this wall after intensely violent riots in the vicinity of Cupar Way on the evening of August 14, 1969, that continued into the following day. The army, anticipating turmoil in Belfast (after the "Battle of the Bogside" in Derry, discussed in Chapter 4), constructed temporary barriers between interface areas, called "knife rests." When violence erupted in the Falls neighborhood (C) (which was more ethnically mixed at that time), it spread throughout the northwestern part of Belfast including the Ardoyne (C) neighborhood, discussed later in this chapter. Police intervention resulted in the death of several of the rioters in "flashpoints" – areas where interactions became particularly fierce (e.g., McKittrick and McVea 2002).

Burning of Bombay Street

These riots triggered a process of forced or self-driven residential segregation in northwest Belfast. The exodus originated along Bombay Street in the Clonard (C)

3 In addition to the impacts of de-industrial urban decay, Belfast's central business district was devastated by WWII bombing raids.

FIGURE 5.2 Detail of Cupar Way peace wall, north Belfast, Northern Ireland.

Source: Photo by author.

neighborhood as rioters and arsonists targeted Catholic residences. As the fires spread and rioting intensified, residents from both communities living near the interface area fled, separating out and creating more ethnonationally homogeneous enclaves (e.g., Ballymurphy and Andersonstown).

Belfast police eventually relinquished control to the army. Despite the presence of the army, which began to patrol the Falls Road area and other contentious areas of northwest Belfast, it had little success ending the riots. Ultimately, the majority of the houses located closest to the ethnonational interface, along Brookfield and Bombay Street, were destroyed by fire. After intimidation, arson, and rioting, 1,820 families (1,505 of which were Catholic) abandoned this previously "mixed" interface area, forging segregated enclaves between the Falls/Clonard (C) and Shankill (P) neighborhoods (Scarman 1972). Throughout the city, over 9,000 houses would also be abandoned or rendered uninhabitable as a result of damage from violence (McKittrick and McVea 2002).

Conflicting national narratives shaped how local residents interpreted the riots, and further alienated the communities. For example, much of the Protestant community believed the rioting was a vital, preemptive strike against an impending IRA uprising (Mulholland 2001). In contrast, many Catholics perceived these clashes as a vicious attack by loyalist mobs as authorities refused to intervene. As politician and Irish civil rights activist Paddy Devlin recounts as a witness to the rioting:

> The sky was red from the burning houses, torched, while the military stood by, not raising a finger.
>
> *(cited in Coogan 2015)*

The traumatic events that began on Bombay Street are memorialized through the Catholic perspective in several local political murals that include the slogan "Never Again!" – reminding residents to remain vigilant against loyalist aggression against their community (Chapter 6 examines these murals).

News of the riots inspired other Catholic inhabitants to create "No-Go" areas, seizing municipal buses, delivery trucks, cars, and scaffolding as makeshift barricades along key transportation routes and residential perimeters.

The Holy Cross "dispute"

Another self-segregating residential exodus occurred in the Ardoyne neighborhood of northwest Belfast. Prior to the Troubles, it was a "mixed," working-class neighborhood, but sectarian violence drove the two communities apart – Catholics now primarily in the south and Protestants in the north in an area called Glenbryn. A peaceline was constructed in Ardoyne to divide the newly segregated communities. However, because of the new residential pattern, Holy Cross Catholic girls' primary school was now located in a subsequently Protestant territory, Glenbryn (see Map 5.2).

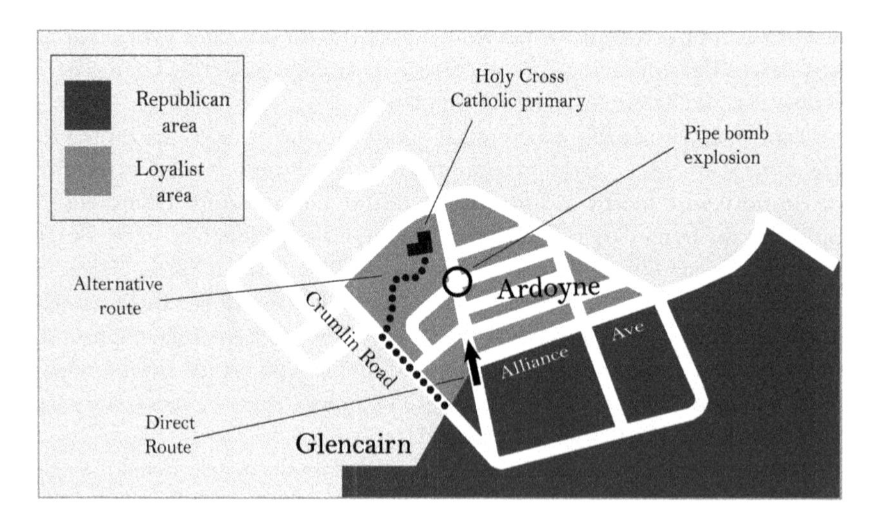

MAP 5.2 Map of Ardoyne/Glencairn neighborhood during the "Holy Cross conflict" [neighborhood are commonly described as either republican (Catholic) or loyalist (Protestant)]

Glenbryn residents, many of whom reported being harassed by republicans, subsequently refused to permit those attending this Catholic school access through Glenbryn. Tensions erupted in June 2001 when Protestants picketed along the walking routes that some Catholic Holy Cross students utilized to arrive at school. Loyalist paramilitaries fomented riots and created blockades to escalate violence. When picketers began to verbally assault and throw stones at Holy Cross students as they walked to school, the RUC was sent to protect the young girls. As the rioting spilled out into surrounding parts of Ardoyne and surrounding northwest enclaves, pipe bombs and other forms of sectarian violence engulfed the city. In the international media, the events were soon equated to the Alabama race riots in the 1960s (e.g., McKittrick and McVea 2002).

When the conflict continued at the start of the next school term in September 2001, the conflict escalated quickly. As loyalists blocked the students' routes, the RUC and the army intervened to create a corridor to escort the Holy Cross schoolgirls, some as young as four years old, and their parents. Picketers, many of whom were forced blocks away from the corridor, threw stones, ball bearings, bottles, and blast bombs at the police and students (Gaffikin et al. 2016). Protestant minister and local interfaith peace advocate, Reverend Bill Shaw (discussed in Chapter 7), volunteered to accompany the Catholic Holy Cross families along their route to school.

The British Secretary of State for Northern Ireland, John Reid, demanded an immediate cessation of the riots, while Richard Haass, the United States' Special Envoy for Northern Ireland, traveled to London to discuss a resolution to the turmoil. Archbishop Desmond Tutu met with local community members of Ardoyne/Glenbryn in an effort to foster peace. However, as the violence persisted for several weeks, the enrollment at Holy Cross halved, and its students began manifesting evidence of psychological trauma (Heatly 2004). When the turmoil finally abetted, tensions remained. In 2003, a pipe bomb attack on the school was unsuccessful but fueled further mistrust between the enclaved communities.

The persistence of Belfast's walls and segregation

While violence has decreased since the Troubles, the size and number of partition walls have increased (Cunningham and Gregory 2014). For example, there were 18 peacelines in Belfast in the 1990s, but the number increased to 88 by 2010 with requests for new constructions despite the government's promise to remove the walls by 2023 (Grattan 2020; McKittrick 2013). Some believe maintaining the peacelines reduces communal violence, particularly at interfaces between contentious enthonational enclaves.[4] Many residents in these contested areas expressed

4 However, research suggests that bodily harm during the Troubles was often concentrated within the heart of an enclave instead of exclusively along contiguous interfaces (Cunningham and Gregory 2014).

they would feel "unsafe" without the walls. Others stated they are not ready to live without walls as many of the underlying conflicts are not yet resolved (e.g., personal interviews conducted June 2017; May 2018; McKittrick 2013).

As one resident of Sandy Row (P), Belfast, explained in a personal interview:

> The walls give us a sense of security. A decade ago, we had all kinds of things thrown at us over the wall. Some of them were really dangerous. We had glass bottles hit one of our windows. Another time a brick almost hit my son. It's hard to forget that. There's still plenty of distrust between the communities here. I just feel safer with the wall.
>
> *(personal interview in Belfast, June 2017)*

Indeed, decades after the 1998 peace agreement, many residents and neighborhood organizations such as Interaction Belfast believe it is necessary to continue monitoring the walls and reporting disruptions.

There are also a growing number of individuals and organizations who challenge the presence of peacelines. Some believe the walls are necessary, but openly scorn their stark and bleak appearances. In some cases, residents have concealed the solid divisions behind newly planted trees or bushes, creating what some call an "environmental barrier." Others, including cross-community programs such as the Duncairn Community Partnership, argue that preservation of peacelines is detrimental to society as they reinforce segregation, isolation, and alienation between the communities (DCP 2017).

Perceptual walls

While segregation intensified during the violence of the Troubles, much of Northern Ireland remains divided along ethnonational lines (e.g., Grattan 2020). Geopolitical narratives of fear and security continue to shape many individuals' classifications of space and people along sectarian partitions. Despite many residents' similar physical appearances (i.e., white), local inhabitants discern ethnonational affiliations through colloquially perceived categorizations such as their name, language/slang, dress, schools, residence, or accent. In essence, many self-segregate through socially constructed bordering practices that produce a shared group identity and the "Other" – dividing and marginalizing communities along invisibly perceived lines of difference (e.g., Dell'Agnese and Szary 2015). As Northern Irish Nobel Laureate, Seamus Heaney, wrote in, *Whatever You Say, Say Nothing* (1975):

> Manoeuvring to find out name and school,
> Subtle discrimination by addresses
> With hardly an exception to the rule
> That Norman, Ken and Sidney signalled Prod[5]
> And Seamus (call me Sean) was sure-fire Pape.

5 "Prod" is a colloquial term for Protestant, and Pape refers to a Catholic.

Social segregation

Ethnonational social segregation shapes daily life for many in Northern Ireland, especially residents in sectarian enclaves where they have little or no contact with individuals from other communities (e.g., Belfast City Council 2017, POP008). For example, employment opportunities are commonly segregated, thereby removing potential interactions or collaborations with members of another community. Public transportation and shopping destinations, as well as pedestrian and driving routes, continue to navigate visible and invisible barriers that exist as residents avoid boundaries of opposing enclaves (Gaffikin et al. 2016). Tracking the movement of residents with GIS data in segregated areas, such as northwest Belfast, revealed reciprocated exclusivity as residents in Belfast's adjoining Shankill (P)/Falls (C) and Ardoyne (C)/Glenbryn (P) only frequent stores, grocers, parks, public transport, and recreational facilities within their own ethnonational community despite geographic proximity to "opposing" facilities (e.g., Nolan 2017).

Northern Ireland's educational system also reinforces ethnonational divisions. Over 93% of school-age children in the region attend government-funded religiously segregated schools (Dowler and Ranjbar 2018). Curriculums within the bifurcated school system often reinforce unchecked sectarian narratives and ethnonational identities, a social mechanism that bolsters social divisions (e.g., Gaffikin et al. 2016). While "integrated schools" exist, entrenched geographies of residential segregation make access to them difficult, and enrollment has not dramatically increased despite the advent of the peace process (e.g., NIC 2020). Predictably, research suggests friendships among children are highly divided along ethnonational lines (e.g., Landow and McBride 2021). "Mixed marriages" between members of the Catholic community and Protestant community are rare. Less than 10% of the marriages in Northern Ireland are "mixed" and usually result in one of the spouses severing ties with their side of the family (e.g., McAuley 2016).

Municipal planning exacerbates the socio-political bordering practice of segregation. In 1969, the Community Relations Commission released a report that argued efforts to bridge communal divisions in Northern Ireland were futile. Subsequently, Northern Irish Housing Executive (NIHE) designed public infrastructure to separate the communities, and by 1999 Belfast social housing was almost 100% segregated (Housing Executive 2016). In 2016, a joint EU and Queens University (Northern Ireland) report on segregated social housing called the strategy "deeply flawed" as it:

> further endorsed the sectarian geographies of many local communities as an unfortunate inevitability of an ethnonationalist contest, rather than calling it what it is: narrow ghettoization, which locked these areas into constricted spaces and visions, and encouraged rivalries over allocation of urban resources.
>
> *(Gaffikin et al. 2016, 78)*

After many such critiques, efforts to develop "mixed" housing within segregated and boundary areas between two contiguous enclaves have had little success. For example, the Girdwood Community Hub straddles two segregated enclaves in north Belfast. Its development plans originally included "mixed" housing as part of its larger comprehensive community design. However, after many local residents expressed fear that integrated housing would provoke violence and few applied to live in the proposed housing structures, Girdwood organizers removed housing and only built a shared recreational facility. During a personal interview with residents near the Girdwood Community Centre in Belfast, one adult male resident explained:

> We just live separately here. Always have, always will. It's how we stay safe. They don't come here and we don't go over there [pointing past a peaceline]. We stay away. [When I asked the interviewee if they knowingly met someone from "over there, especially since the neighborhood is located so close to his own" he responded] No, never.
>
> *(personal interview in Belfast, 2017)*

Personal interviews with a group of 14–15-year-old students from the Falls neighborhood (C) reveal that the majority have never met a Protestant. Similarly, a group of similar-aged students in the contiguous Shankill neighborhood (P) did not know any Catholics (personal interviews, Belfast 2016). As two teenage interviewees from adjacent neighborhoods explained in interviews:

> I don't know any Catholics. No one here does either. It's just the way it is here in Belfast. [I asked if he ever went into a Catholic neighborhood]. No, I don't go over there, I'd get hurt if I did, so I stay away.
>
> *(personal interview in Shankill [Belfast], 2016)*

> They're not like us. They're British. They're Protestants . . . We stay away from them, which is good. You know we go to different schools here in Belfast, right? You see, it's better if we stay apart.
>
> *(personal interview in Clonard [Belfast], 2016)*

These bifurcated perceptions are common. For example, during one of the study-abroad trips I led to Northern Ireland, we visited schools in north Belfast. The local students pleasantly greeted us and introduced themselves. They then asked my students their names and, eventually, if they were Protestant or Catholic. All my students responded, but one of my students, who was Jewish, identified herself accordingly. The local students greeted her enthusiastically but then waited expectantly. Finally, one of the local students responded, "Welcome! But, we want to know, are you a Protestant Jew or a Catholic Jew?" (personal interview in Belfast, 2014).

Physical isolation as a result of socio-political bordering contributes to the protracted segregation in Northern Ireland. "Contact Hypothesis" – a theory of research utilized in various fields including psychology, geography, sociology, and criminology – suggests that one of the most effective ways to encourage tolerance between two conflicting groups is through constructive interactions, especially if the two groups can work together for a common goal (e.g., Enright and Rothschild 2015). However, the opposite, complete separation, can prolong and reinforce rivalries. Without meaningful interactions with "counterparts," sectarian narratives of hatred and mistrust of "the Other" are commonly fostered and strengthened.

Many in enclaves believe residential segregation is necessary for security and self-preservation. For adults, the desire to divide along sectarian lines commonly resulted from a negative personal experience that most frequently transpired during the Troubles. However, accounts of these experiences can shape younger generations' perceptions of "rival" communities, even if they have not met a member of the "opposition." Indeed, interviews with children in Northern Ireland reveal hatred or fear of the "Other" as a result of events that occurred before their birth, possibly reflecting narratives about such events are often recounted with vehemence. The power of collective memory, which in enclaves commonly emphasizes traumatic past events and mistrust, can greatly influence individuals' perceptions and assumptions of rival enclaves (Enright and Rothschild 2015). Therefore, it is not surprising that studies suggest since the 1990s, segregation has increased in these contested areas (e.g., Paolini et al. 2016).

Segregated enclaves are often marked territorially with flags, graffiti, murals, and street-curb paintings (Nolan and Bryan 2016). These symbols reinforce difference, foster exclusionary national narratives, and discourage opportunities for meaningful contact with the "Other." Despite the presence of minorities, diversity, and dissent, the presumed reality in these neighborhoods is powerfully focused on Catholic-Protestant binary categorizations.

Flags as territorial markers

Political flags are spatial signifiers. Public display of these flags can serve as a powerful symbol of identity, territorial marking, and sovereignty. Flagging is also a common element of statecraft, a visual demonstration in which a territorial claim is made legible (Dodds 2010). These images are a shorthand that articulate certain meanings and values and can foster a sense of community or spark political disagreements. While flying a state flag may appear to be an uncontentious action for some living outside of this region, this is not true in Northern Ireland. Fueled by a tumultuous sectarian history, controversies over political flags (e.g., Union flag or Irish tricolor), paramilitary flags, and sectarian banners in Northern Ireland are not uncommon.

For example, while many unionists (and particularly loyalists) revere the UK's Union flag, the "Union Jack," it is widely scorned by many nationalists and republicans in Northern Ireland because they interpret the flag as a material representation of British colonialization of Ireland. In contrast, nationalists or republicans

FIGURE 5.3 Irish "tricolor" flag.

FIGURE 5.4 UK "Union Jack" flag.

may display the Republic of Ireland's "tricolor" flag to signify their Irish identity and culture, despite residing in the UK (e.g., McDowell, Braniff, and Murphy 2017) (see Figures 5.3 and 5.4).

The significance of flags

The symbolic importance of political flags in Northern Ireland increased after the failed 1994 paramilitary ceasefire, as their presence commonly signified continued sectarian support (Nolan and Bryan 2016). Often flags appear on personal residences and lampposts in enclaved neighborhoods as assertive symbols that territorially mark a residence or enclaves' ethnonational affiliations.

Research on the employment of sectarian flags in Northern Ireland suggests loyalists utilize these symbols more frequently than republicans. For example, during marching season (discussed in Chapter 6), researchers annually record close to 4,000 flags displayed along streets on lampposts, phone lines, and houses along parade routes. These are not only predominantly loyalist or unionist political parties' flags but also loyalist paramilitary UVF and YCV (Young Citizen Volunteers) organizations.[6] While regulations stipulate the length of time that political flags can be displayed in Northern Ireland, they are not uniformly enforced. For example, while the "marching season" spans April–August, many union/loyalist flags remain into September. Research on the impact of these "residual" flags suggests that their continued presence is unpopular (i.e., 80% objected to their presence) and specifically discouraged certain individuals from shopping in "flag areas," as they felt unwelcome or unsafe in the space (Nolan and Bryan 2016).

Many Catholics also express aversion to public displays of the Union flag, which they interpret as a symbol of Protestant domination over Catholics in Northern Ireland. Others explain how the presence of the Union flag and/or loyalist paramilitary flags in a neighborhood makes them feel threatened and unsafe. Conversely, Protestants report feeling endangered in areas displaying republican flags or the Republic's tricolor (personal interviews conducted in 2014; 2015; 2018). Due to the provocative nature of flags, there have been some attempts to remove the more contentious ones throughout Northern Ireland.

The 1989 Code of Practice mandated that in Northern Ireland all flags and national insignia be removed from the public workplace, including the Union flag. However, in 2000, the Secretary of State for Northern Ireland approved the "Flags and Emblems Act" that provided parameters for flags on government buildings. Crucially, it permitted the 26 district councils to create their own flag policies. In 2012, Belfast City Council (BCC) elected to reduce the number of days it displayed the Union flag at City Hall from 365 to 18. Its new flag legislation ignited region-wide protests, and the subsequent controversy over the public display of the Union flag remains one of the most intractable concerns in Northern Ireland.

Loyalists' response to the reduction of days the flag is publicly displayed included riots, picketing, and roadblocks on the evening of the 2012 vote. This was followed by 4 months of street protests, political squabbling, and a short-term rise of "flag-protest political parties." During these months of political turmoil, police reported 2,980 physical incidents and 55,521 public demonstrations (Nolan and Bryan 2016). Some of these protests resulted in physical injuries to citizens and 160 police officers (Nolan et al. 2014).

According to political geographers in the region, political flags hold great significance in Northern Ireland:

> Outside observers express disappointment that the peace process has still not managed to resolve what, on the face of it, appears to be a minor issue about flags, but which manage to fuel sustained bout of civil unrest.
>
> *(Nolan and Bryan 2016, 3)*

6 The number of paramilitary flags publicly displayed has decreased over the last two decades.

For example, many of the protestors interpreted the reduction of flag days as "a symbolic battle that they believed had 'far reaching consequences' . . . indicative of both a lack of respect for their culture and identity, as well as the 'loss' of City Hall" (McDowell, Braniff, and Murphy 2017, 197). The protests revealed how many in sectarian communities felt unrepresented by the political leadership and flag legislation in the region. However, the complicated and emotional nature of debates over flags manifested in a variety of multifaceted, and sometimes conflicting manners. For example, not all who supported the daily display of the Union flag at City Hall lack consideration for Catholic/nationalist/republican interpretation of the flag. Instead, some individuals who actively participate in or support cross-community projects simply believed it was appropriate to fly the state flag in a region of the UK.

Protests and counter-protests became violent as the turmoil continued into Protestant marching season in July. During this time, tens of thousands of Union flags appeared in many Protestant enclaves on houses or street lampposts. As the ensuing controversy triggered a deterioration in community relations in Belfast, skepticism over local authorities' ability to appropriately and effectively resolve the conflict increased (Nolan et al. 2014, 102).

The following year, a multi-party task force was formed to identify common ground for flag usage and policies in Northern Ireland. However, it failed to reach a consensus. For some Protestants, public and legislative challenges to displaying the Union flag – which they interpret as a symbol of their history, identity, and culture – are difficult to accept. Subsequently, many are suspicious of discussions of a "shared future" and cross-community reconciliation efforts (Nolan and Bryan 2016). Indeed, the contentious nature of displaying political flags in Northern Ireland underscores how flags can be highly contested and symbolic-territorial claim to space (Nolan et al. 2014). As the flag task force 2014 report stated:

> Without a larger consensus on the place of Britishness and Irishness – for which there must be a protected place alongside other identities, national or otherwise, represented in our society – we could not reach a common position on the flying of flags and the display of other emblems, which are in fact manifestations of those identities.
>
> *(Haass and O'Sullivan 2014)*

The price of a divided society

The number of peacelines in Northern Ireland has increased since the 1998 peace agreement. Enclaves are considerably more homogenous than before the Troubles, as peacelines reinforce a form of sectarian apartheid. While government-funded projects foster cross-community interaction and reconciliation, removal of walls remains relatively rare (Cosstick 2017). In 2020, 116 peacelines remain, many of which are not clearly demarcated on tourist maps (Grattan 2020). Perhaps this reveals intentional efforts to downplay or ignore their presence and the region's protracted ethnonational divisions.

Over the years, many have examined the cost of the partition walls in Northern Ireland (and comparably divided societies throughout the world) and asked – What are the impacts of maintaining segregated communities along ethnic partitions? Evidence of residential abandonment and disrepair of areas immediately adjacent to the walls are common. Economic growth for these areas is limited and underpins divided residential geographies as housing markets within enclaves remain cemented along ethnonational lines. Urban planners and municipal leaders perpetuate the dereliction, as many lack strong initiatives to remove the barriers or invest in the enclaved communities.

The construction of walls also alters transportation and walking routes to hospitals, employment, shopping centers, and schools. It impedes cross-community activities in public areas, including public parks, some of which have been physically severed by peacelines. Segregated schools and employment make positive cross-communal interactions difficult and rare for residents in sectarian enclaves. If friendships across communal boundaries exist, many admit to avoiding discussions of ethnonationalism amid "mixed" company (personal interviews in Derry, Coleraine, Belfast, 2016; 2017).

The presence of peacelines signifies not only the perceived necessity for "protection" but also a desire to divide – to remain separate and isolated. Some opponents of segregation argue that while walls may reduce violent conflict in the short term, their permanency isolates residents and maintains ethnonational divisions. Some critics have challenged the term "peaceline," suggesting that since they do not produce *peace*, the term is instead a "palatable euphemism for 'segregation walls'" (Gaffikin et al. 2016, 1).

The walls also imprison residents within their enclaves, enabling and exacerbating hostilities within and beyond divided communities' social borders. As research demonstrates, many residents feel trapped within their neighborhoods and unable to leave for safety reasons. Others fear physical or verbal attacks if they attempt to venture outside the walls. Indeed, the peacelines symbolically and physically solidify divisions between rival communities, planting seeds of future forms of xenophobia and conflict. As Contact Hypothesis suggests, these segregated communities will continue to divide along sectarian lines without opportunities for positive interaction.

A legacy of violence, sectarianism, and segregation

Partition walls and political flags are symbolic signboards forged across Northern Ireland's landscape. Their impact as divisive and provocative spatial displays of territoriality and contestation is hardly benign. Indeed, as described in this chapter, the presence of peacelines exacerbates sectarian conflict and impedes reconciliation efforts. The geographic patterns of division and detachment have been protracted in many of the enclaves. Flags territorially mark space along ethnonational lines, alienate those identified with a different community, and perpetuate divisions.

In 2014, Deputy First Minister of Northern Ireland, Martin McGuiness, acknowledged that the continued erosion of political relations between ethnonational communities threatens the peace process. While violence has decreased since the 1998 peace agreement, vestiges remain – particularly in these segregated communities. The following chapter continues our examination of the Troubles and ethnonational territorialized identities in Northern Ireland. Chapter 6 interrogates how space is (re)affirmed through the movement of bodies through the act of political parading and the creation of political murals, to offer insight into geographic practices that transform parading bodies and murals into ethnonational markers of difference.

References

Belfast City Council. 2017. "Local Development Plan 2020–2032." POP008. www.belfastcity.gov.uk/getmedia/1d2a917d-3d3f-4f74-a29e-c8d360a0c83d/POP008_TP-Pop.pdf.

Belfast Housing Executive. 2016. "Housing Executive's Community Cohesion Strategy 2015–2020."

Coogan, Tim P. 2015. *The Troubles: Ireland's Ordeal 1966–1995 and the Search for Peace*. Michigan: Head of Zeus Ltd.

Cosstick, Vicky. 2017. "Belfast Peace Wall Removal Opens up Fresh Social Visits." *Irish Times*, July 21, 2017. www.irishtimes.com/news/ireland/irish-news/belfast-peace-wall-removal-opens-up-fresh-social-vistas-1.3161777.

Cunningham, Niall, and Ian Gregory. 2014. "Hard to Miss, Easy to Blame? Peacelines, Interfaces and Political Deaths in Belfast during the Troubles." *Political Geography* 40 (May): 64–78. https://doi.org/10.1016/j.polgeo.2014.02.004.

DCP. 2017. "Duncairn Community Partnership 2017." www.duncairn.com/events-for-all.php#pde.

Dell'Agnese, Elena, and Anne-Laure Amilhat Szary. 2015. "Borderscapes: From Border Landscapes to Border Aesthetics." *Geopolitics* 20 (1): 4–13.

Dodds, Klaus. 2010. "Flag Planting and Finger Pointing: The Law of the Sea, the Arctic and the Political Geographies of the Outer Continental Shelf." *Political Geography* 29 (2): 63–73.

Dowler, Lorraine, and A. Marie Ranjbar. 2018. "Praxis in the City: Care and (Re) Injury in Belfast and Orumiyeh." *Annals of the American Association of Geographers* 108 (2): 434–44.

Enright, Robert, and Babette Rothschild. 2015. *8 Keys to Forgiveness (8 Keys to Mental Health)*. New York, NY: WW Norton & Company.

Gaffikin, Frank, Chris Karelse, Mike Morrissey, Clare Mulholland, and Ken Sterrett. 2016. *Making Space for Each Other: Civic Place-Making in a Divided Society*. Belfast: Queen's University Belfast.

Grattan, Steven. 2020. "Northern Ireland Still Divided by Peace Walls 20 Years after Conflict." *The World*, January 14, 2020. www.pri.org/stories/2020-01-14/northern-ireland-still-divided-peace-walls-20-years-after-conflict.

Gregory, Ian N., Niall A. Cunningham, Paul S. Ell, Christopher D. Lloyd, and Ian G. Shuttleworth. 2013. *Troubled Geographies: A Spatial History of Religion and Society in Ireland*. Bloomington: Indiana University Press.

Haass, R., and M. O'Sullivan. 2014. *Proposed Agreement": An Agreement among the Parties of the Northern Ireland Executive on Parades, Select Commemorations, and Related Protests; Flags and Emblems; and Contending with the Past*. Belfast: www. northernireland.gov.uk/haass. pdf.

Hayes, Bernadette, and Ian McAllister. 2013. *Conflict to Peace: Politics and Society in Northern Ireland over Half a Century*. Manchester: Manchester University Press.

Heaney, Seamus. 1975. *Whatever You Say, Say Nothing*. London: Faber & Faber.

Heatly, Colm. 2004. *Interface: Flashpoints in Northern Ireland*. Belfast: Lagan Press.

Kelters, Seamus. 2013. "History – The Troubles – Violence." *BBC*, 2013. www.bbc.co.uk/history/topics/troubles_violence.

Landow, Charles, and James McBride. 2021. "Moving Past the Troubles: The Future of Northern Ireland Peace." *Council on Foreign Relations*, 2021. www.cfr.org/backgrounder/moving-past-troubles-future-northern-ireland-peace.

McAuley, Jim. 2016. "Memory and Belonging in Ulster Loyalist Identity." *Irish Political Studies* 31 (1): 122–38.

McDowell, Sara, Máire Braniff, and Joanne Murphy. 2017. "Zero-Sum Politics in Contested Spaces: The Unintended Consequences of Legislative Peacebuilding in Northern Ireland." *Political Geography* 61: 193–202.

McKittrick, David. 2013. " 'Taking Belfast's Peace Walls down Solves Nothing. They Should Be Higher:' Protestants and Catholics Agree Time Is Not Right to Dismantle Barriers." *The Independent*, 2013.

McKittrick, David, and David McVea. 2002. *Making Sense of the Troubles: A History of the Northern Ireland Conflict*. Chicago: New Amsterdam Books.

Mesev, Victor, Peter Shirlow, and Joni Downs. 2009. "The Geography of Conflict and Death in Belfast, Northern Ireland." *Annals of the Association of American Geographers* 99 (5): 893–903. https://doi.org/10.1080/00045600903260556.

Mulholland, Marc. 2001. *The Longest War: Northern Ireland's Troubled History*. Oxford: Oxford University Press.

Murphy, Dervla. 1979. *A Place Apart*. Waterford: Irish Book Center.

NIC. 2020. "Northern Ireland Council for Integrated Education." www.nicie.org/wp-content/uploads/2020/07/NICIE-Directors-Report-and-Financial-Statements-for-year-end-31.3.2020-Final.pdf.

NISRA. 2019. "Northern Ireland Statistics and Research Agency (NISRA) – Northern Ireland Housing Statistics." www.communities-ni.gov.uk/system/files/publications/communities/ni-housingstats-18-19-full-copy.PDF.

Nolan, Paul. 2017. "Two Tribes: A Divided Northern Ireland." *The Irish Times*, April 1, 2017. www.irishtimes.com/news/ireland/irish-news/two-tribes-a-divided-northern-ireland-1.3030921.

Nolan, Paul, and Dominic Bryan. 2016. *Flags: Towards a New Understanding*. Belfast: Institute of Irish Studies, Queens University.

Nolan, Paul, Dominic Bryan, Clare Dwyer, Katy Hayward, Katy Radford, and Peter Shirlow. 2014. *The Flag Dispute: Anatomy of a Protest*. Belfast: Queen's University Belfast.

Paolini, Stefania, Miles Hewstone, Alberto Voci, Jake Harwood, and Ed Cairns. 2016. "Intergroup Contact and the Promotion of Intergroup Harmony: The Influence of Intergroup Emotions." In *Social Identities*, 209–38. East Sussex: Psychology Press Ltd.

Scarman, Leslie. 1972. *Violence and Civil Disturbances in Northern Ireland in 1969: Report of Tribunal of Inquiry*. Belfast: H.M.S.

6

ETHNONATIONAL SPATIAL STRATEGIES AND TRANSGRESSIONS

Political parades and murals

The act of political parading in Northern Ireland underscores competition for validation and belonging between perceived "rival" communities. As a parade participant explained:

> Marching is our tradition. It is a very important part of our Protestant community's history. It's our culture and we need to keep marching. We need to be seen and heard. If we don't, the others will take us over.
>
> *(personal interview with a marching participant, Belfast, July 2017)*

As these parades are politically charged, when a route transgresses an ethnonational boundary, it often results in turmoil. Like parades, political murals serve as a medium through which a community displays territorial claims or selective memories of historical events. While these artistic statements were often designed to broadcast ethnonational messages or rally supporters for a sectarian cause, recent efforts have focused on reconciliation and cross-communal healing.

The previous chapter investigated segregation, peacelines, and flagging to examine how landscape is employed to symbolically mark and reinforce sectarian territorial and social divisions in Northern Ireland. This chapter builds on the previous case study to interrogate how ethnonationalism is spatially mobilized through political parading and the creation and strategic location of political murals. Historically, these murals were created in concurrence with political parades to imbue the route with greater meaning and significance. Over time, the two (i.e., parades and murals) evolved into separate forms of geopolitical practice in Northern Ireland.

The first part of this chapter explores the evolution of political parades in Northern Ireland and the organizations associated with their production. The following section introduces murals as a symbolic medium for inscribing geopolitical declarations and territorial claims through the local landscape. This chapter then

DOI: 10.4324/9781003141167-6

discusses the history of murals as well as their significance within Northern Irish society. This is followed by examples of unionist/loyalist and nationalist/republican murals as well as common themes identified in each community. Finally, this chapter explores recent efforts to modify aggressive sectarian murals and parades within this divided society.

Parades: marking and moving through territory

While Northern Ireland's peacelines are visceral barriers that mark and enclose a community, the boundaries between enclaves in contested areas are not always divided by physical constructions. Imaginative boundaries between communities exist in locals' minds and transit routes, functioning in the same manner as peacelines. For example, many rural communities, particularly in the center of Ulster, remain segregated without the presence of physical walls. Instead, boundaries of belonging are ceremonially marked and reaffirmed through embodied spatial practices such as the movement of bodies across space through the act of parading, which mark and underpin territoriality and perceived shared ethnonational identity.

Indeed, parading is a ritual practice that is primarily about geography – the power to command and exert control over space and foster a sense of belonging (Hagen and Ostergren 2006). Because the act of parading in Northern Ireland is steeped in historical and political memory, it has become a lightning rod for conflict, particularly within enclaves as conflicting and adversarial memories publicly compete for validation and recognition across space (McDowell, Braniff, and Murphy 2015). This is exceptionally true if a parade route encounters or transgresses a physical or invisible boundary that divides the entrenched communities.

There are a variety of parades in Northern Ireland, some transcend ethnonational divides such as pride parades and civil rights marches, while others are highly sectarian in nature. Parades celebrating significant political events or religious holidays have been an important part of communal traditions in Ireland since the fifth century. However, political parades that commemorate only one of Northern Ireland's ethnonational histories are highly contentious (e.g., Jarman 2020).

For example, Protestant parading predates the establishment of Northern Ireland (i.e., the earliest notable display recorded was in 1660). Critics suggest these parades reflect territorial efforts to control and defend their community on an island on which they are the demographic minority (e.g., McAuley 2016). In preparation for these parades, routes are adorned with Union flags, political and paramilitary banners, red-white-and-blue bunting (Union flag), painted street curbs, and political murals. This communal embodied practice commemorates their past and reaffirms their commitment to the crown (e.g., Battle of the Boyne, Siege of Derry, and the Battle of the Somme, discussed later in this chapter). It also serves as a mechanism through which territorial claims and performances of "Protestant unity" are performed.

While parading has a long history within much of the Protestant community, Catholics also participate in commemorative and religious parading, albeit to a smaller degree. It is commonly acknowledged that parading does not hold the same significance for the Catholics as Protestants who more frequently employ the spatialized practice to foster, maintain, and protect their ethnonational sense of belonging (e.g., Rolston 2012). According to municipal reports, the ratio of recorded Protestant/loyalist to Catholic/republican parades in Northern Ireland is about nine to one, with a sharp increase in number and ferocity of loyalist parades after the Anglo-Irish Agreement (AIA) in 1985 (Nolan et al. 2014). Indeed, as many within the Protestant, and particularly the loyalist community, felt threatened by the AIA, they organized local parades to publicly reinforce their British identity, heritage, and dedication to its union with the UK. The marches also corresponded with an increase in militaristic loyalist murals, which is discussed later in this chapter.

Less frequently, Catholic/Irish/republican parades organized by the Ancient Order of Hibernians, Irish National Foresters, and some republican groups utilized this spatial form of civic and cultural remembrance. However, the practice of parading was only highlighted within the Catholic community during the latter part of the Troubles. Theirs emphasized celebrating their ethnonational history and protesting the systemic discrimination Catholics faced in Northern Ireland (Jarman 2020).

Protestant parade organizations and the "marching season"

Several institutional organizations facilitate Protestant "Orange" parades. For example, the Loyal Orange Institution or "Orange Order" (discussed in Chapter 3) is the Protestant fraternal organization primarily responsible for the organization and execution of thousands of parades throughout Northern Ireland, the Republic of Ireland, Scotland, and the United States. This institution was named in honor of Protestant King William III (i.e., William of Orange or "King Billy").

During the Williamite-Jacobite War in Ireland (1688–1691), Dutch Protestant challenger William III of Orange vanquished Catholic King James II of England, Ireland, and Scotland (i.e., James VII of Scotland at the Battle of the Boyne, 1690). This war, which included the Siege of Derry (1689), was fought between supporters of the Catholic King James II and those who backed his successor, Protestant William III. His victory over Catholic King James II secured Protestant control of Ireland. Many within the Protestant community in Northern Ireland colloquially refer to him as "King Billy" and include his depiction in many of their political murals. These battles and William of Orange are described later in this chapter.

The Loyal Orange Institution or the Orange Order endeavors to maintain the Protestant Ascendancy and defend their civil and religious liberties; the Order accepts only Protestants within its membership. Adherents wear bold orange sashes while parading and are commonly known as "Orangemen." The Orange Order and similar organizations such as the Royal Black Institution, Apprentice Boys of

Derry, and Association of Loyal Orangewomen of Ireland collectively organize parades memorializing historical events in Protestant Ireland's history throughout the parade season. This season begins at Easter and culminates with a particularly concentrated period of parading in July–August.[1] Since many of these commemorations traditionally occur in the summer, this period is commonly called the "marching season." There are notable differences between rural and urban parading styles. For example, rural celebrations are often more religiously focused, stately affairs that often include biblical imagery in their parades. In contrast, urban parades are louder and secular in nature, often manifested as a military drum-band style marching (e.g., Parades Commission 2020; Jarman 2020).

The largest celebration during the marching season is held annually on July 12. Colloquially referred to as "the Twelfth," this celebration commemorates the William of Orange's victory during the Battle of the Boyne. While the largest and most famous "Twelfth parade" is celebrated in Belfast, several similar fraternal societies also hold feeder parades throughout Northern Ireland and some Ulster towns in the Republic.

Bonfires

In addition to parading, massive bonfires[2] commonly accompany parades, especially for the Twelfth, in Protestant/loyalist neighborhoods. Constructed weeks or months in advance from wooden pallets and wood scraps, these "towers" regularly reach heights of over two stories high and are lit on the eve before the day's events (see Figure 6.1).

As they burn, collapsing materials can ignite external fires that damage adjacent houses and buildings.[3] In addition to their potential to result in physical harm or property damage, these bonfires are controversial for many within the Catholic community, who attribute an increase in hate crimes committed against Catholic residents to the bonfire celebrations (e.g., Jarman 2020). Despite the controversies, those who annually participate in their construction and immolation consider lighting bonfires during the Twelfth as their civic right and fiercely defend this practice. In 2017, when Belfast City Council ruled to limit the height of these constructions for safety reasons, certain members of the loyalist community, the DUP (Democratic Unionist Party), and PUP (Progressive Unionist Party) claimed it was an unwarranted government attack on the sanctity of the Twelfth celebrations (Belfast City Council 2017).

Political parades during the Troubles were particularly controversial. For example, after a particularly violent period in Belfast in July 1970, Stormont banned

1 For a description of each organization and key parades, see Parades Commission for Northern Ireland Annual Report 2020; Jarman 2020.

2 Bonfires also occur in nationalists' communities, particularly on August 15, to commemorate anti-Internment and a Catholic feast day. These bonfires are also extremely controversial.

3 For example, see the Sandy Row fire of 2017.

FIGURE 6.1 Preparing for the "Twelfth" bonfire, Belfast.

Source: Photo by author.

parading for 3 days. However, the Grand Master of the Orange Institution criticized the decision to ban parades on the eve of the Twelfth. He warned that denying citizens the right to march would result in ferocious civil and political upheaval. Ultimately, Twelfth parades in Belfast were reinstated,[4] but conflicts over the parades continued.

Conflicting significance(s) of parading

The significance of these parades is highly contested. For example, while various interpretations exist within any community, many Protestants perceive Catholic parades as celebrations of republican paramilitary violence as well as their desire to abolish Northern Ireland and govern the island of Ireland. Many Protestants also considered civil rights parades during the Troubles ethnonational in nature, despite being both Catholic and Protestant-led. To their detractors, these parades

4 However, the Apprentice Boys of Derry's parade, which commemorates the Siege of Derry in 1689, was canceled as part of a rare 6-month universal ban on parading that was implemented shortly after the tumult in Belfast (Jarman 2020).

reinforced Protestant's minority status on the island and further bolstered a "siege mentality."

While many Protestants believe parading is their right, responsibility, and part of how they maintain their identity and heritage, many Catholics interpret Protestant parades as symbolic and material verification of Catholic's second-class status within Northern Ireland. Thus, both communities perceive these political parades as ethnonational and – particularly for Catholics –threatening. Indeed, Catholics report feeling intimidated, "out of place," or even threatened by the presence of Protestant parades and decorations that adorn their parade routes. To some, the celebratory decorations and revelries of historical events, such as the Battle of the Boyne, are the material manifestation of the Protestant community flaunting their spatial, political, and economic domination in Northern Ireland.

In contrast, others perceive these parades as a mechanism by which the Protestant community:

> reveals that it has lost control of symbolically significant space, seeks to demonstrate an ongoing attachment to critical places . . . the parades narrate their experience in the town of Derry/Londonderry as a victory, despite circumstantial evidence which suggests otherwise. The ability to claim victory through parading provides members of the Apprentice Boys organization with a raison d'etre and serves in place of an aggressive agenda to regain control of territorial icons.
>
> *(Cohen 2007, 951)*

Parade routes have also been a source of great contention. Contending that marching is their right, Protestant/unionists/loyalists argue that their parades should not have any limitations, including its path and desired route. However, when parades border or transgress boundaries between ethnonational communities, riots and violence commonly ensued. Historically, this was most frequent when loyalist parades entered or past by a nationalist/republican enclave.[5] For example, residents of "rival" ethnonational communities have violently reacted to the presence of a "trespassing mob" pass alongside their neighborhood. Conversely, some parade participants have instituted violence, arson, and destruction of property against other ethnic enclaves along their parade path.

Attempts to curtail conflict

Many residents attest to feeling threatened by the presence of the controversial parades. Some have been victimized by physical attacks, projectiles, or other

5 The opposite is controversial as well. For example, when NICRA civil rights marches entered Protestant or "mixed" areas, many Protestants protested, claiming these marches were an invasion of Protestant territory (Bardon 1992).

threatening means. Many businesses, community halls, and schools located along parade routes are preemptively closed and boarded up to reduce any physical damage that may result from riots. Some community programs such as Corrymeela, 174 Trust, or The Ulster Project remove youth from these contentious areas during marching season to protect them from violence and witnessing sectarianism in their neighborhoods (e.g., Corrymeela Magazine 2018).

As a result of mounting pressure, the government commissioned a report to examine the impact of marching season. The study indicated a marked increase in violence coinciding with Orange parades when their routes bordered or entered Catholic/nationalist/republican enclaves. In an effort to manage some of the conflict regarding parades, the state established the Parades Commission in 1998 with authority to re-route, restrict, or ban any parades within Northern Ireland (e.g., Parades Commission 2020). However, many loyalists do not recognize the authority of this organization, arguing it is their civil duty and civic right to parade. More recently, the Orange Order has expressed a willingness to discuss matters relating to parading with the Parades Commission. Despite recent efforts, parading remains a contentious subject in Northern Ireland and frequently sparks riots and violence (e.g., see PA 2018).

While marching is a temporal performance of power and territorial control over space, political murals, which often are paired with marching season and then remain on display indefinitely, also serve as a territorial signifier. As the next section demonstrates, within Northern Ireland's contested landscape, murals act to territorially demarcate the boundaries of a community, simultaneously reinforcing a desire to forge a unified ethnonational identity and warn "Others" they do not belong in this space.

Murals as symbolic and spatial demarcations of ethnonational territoriality

Ethnonational territorial claims are spatially declared in contested areas through the creation of large, colorful political murals. These are prominently painted on the sides of apartment buildings, house gables, and intermittently on peacelines in highly visible areas of an enclave. These murals are usually two or three stories in height and are frequently concentrated in areas where adjacent rival ethnonational communities border one another. For example, the greatest cluster of political murals is located in Belfast, with over 100 within its city limits.

The presence of these spatially provocative political and cultural landscapes works to foster a sense of territorial belonging and to help delineate territorial boundaries among ethnonational enclaves. Geographers often conceptualize symbolic landscapes as complex instruments of socio-political power that reflect processes of change, social conflict, or resistance (e.g., Johnson 2002; Dempsey 2012). They also reveal efforts to influence geopolitical imaginations of place. Indeed, the relationship between the symbolic landscape and society is important as it can function as a dynamic medium through which society is negotiated and principal forms of socio-political messages are displayed. The meaning and significance of these murals is a dynamic and evolving process through which individuals (de)code and contest their meaning and significance. As the power to control and shape landscapes is not evenly distributed throughout society, the subsequent socio-political implications reveal much about local perceptions, traditions, and underlying tensions that may exist within a community.

These murals are not "innocent," and the intended purpose and meaning of these "representative" projections often generate various forms of contestation within the community. Thus, the public will not always agree with the intended ideological plan of these symbolically charged murals, particularly during periods of political or social transformations. Indeed, many political murals have been the focus of scorn, graffiti, or defacement (e.g., Jarman 2020). Such critical interpretations offer the opportunity to study forms of resistance or subversion that may exist in relation to official meanings or conceptions. For example, murals' bold statements, integrated within an emotive form of vernacular and political art, present selected messages to the public. Subsequently, political parties, politicians, or paramilitary organizations have expressly commissioned murals as a form of promotional propaganda. These promotional pieces are commonly geographically situated in close proximity to significant locations, such as the corresponding party's headquarters.

While these iconic murals are visible throughout Northern Ireland, they are concentrated in less affluent and highly sectarian segregated urban areas. Because these murals are relatively cheap to create and maintain, they can serve as promotional billboards for a local community. Indeed, in more recent years, the locations of the most elaborate and poignant murals have been strategically placed along central thoroughfares, intersections, or high-traffic areas for increased visibility (see Map 6.1). In this way, many murals can also be interpreted as a medium through which the voice of the disenfranchised is made public.

To ceremonially empower a mural, a promotional speech by a local dignitary or social elite before a crowd and the media often preceded its public unveiling.[6] Such political performances are designed to provide "weight" to a mural's significance in a community. In this way, these symbols, mottos, and historical reminders are endowed with great territorial meaning. They become a visual form of a community's self-expression, responses to perceived threats, and nostalgic longings for the past.

These murals are also designed to be extremely accessible to a broad audience. These highly evocative and impactful murals hold a variety of powerful meanings for observers, tourists, and local residents. In recent years, these images have traveled far beyond the local neighborhood through media coverage, books, and internet blogs discussing the murals as well as through "conflict tourism" as tourists

6 However, these ritual practices became less common with the loss of Northern Ireland's parliament (i.e., the support from the local government) under the period of London's "Direct Rule."

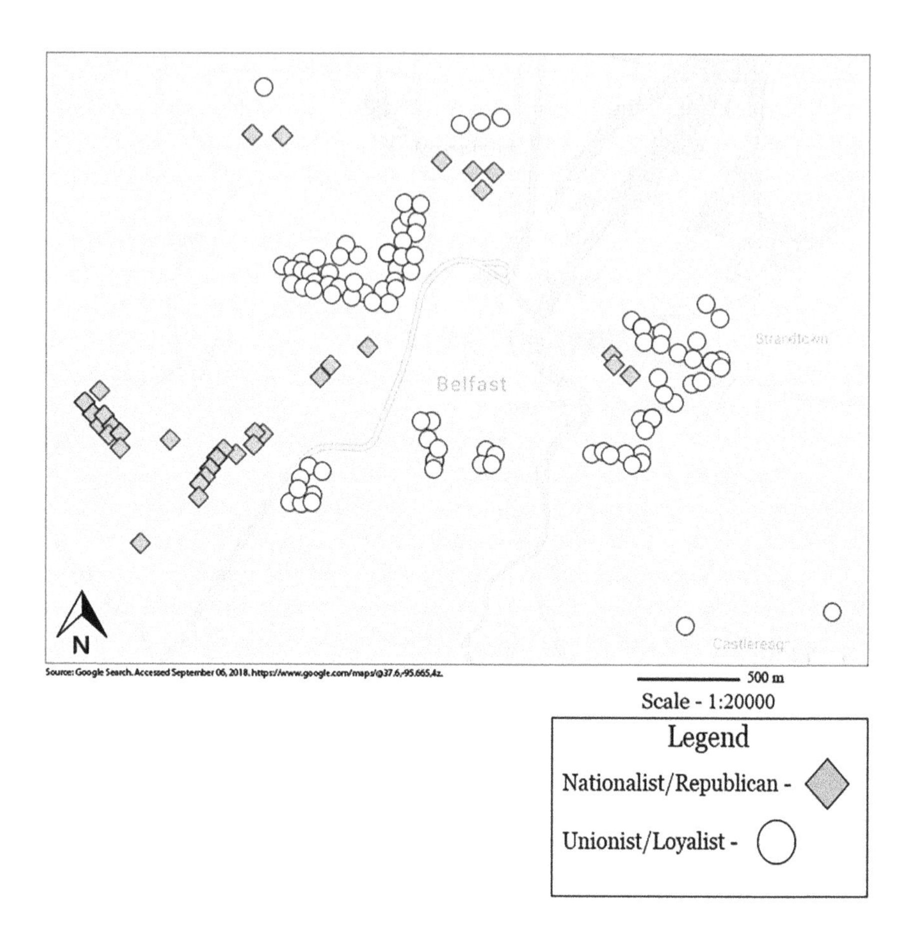

MAP 6.1 Map of location of many political murals in Belfast

Source: Map by Max Martin.

travel to visit the murals. Widely acknowledged as one of the most recognized elements of the political landscape of Northern Ireland, these murals subjectively paint competing interpretations of history, identity, and the past.

A brief history of murals in Northern Ireland

As previously stated, the first political murals in Northern Ireland were created during the first decade of the twentieth century to decorate Orange parade routes, particularly in anticipation of the Twelfth. The first murals in Northern Ireland were historical in nature, often featuring William of Orange in the Battle of the Boyne (1690). In addition to murals, locals commonly painted street curbs on the route (often orange or colors of the Union Jack) and hung flags and/or bunting along streets. Since these decorations were removed shortly after the event, murals provided a more permanent means to mark territory and reinforce the shared history of an ethnonationally identified community.

Catholics in Northern Ireland did not enjoy the same power or representation in public and political matters as Protestants in the region. For example, the state, the army, municipal police, and loyalist paramilitaries commonly defaced or destroyed Catholic/pro-republican murals (with paint bombs or more destructive methods). Muralists and residents who lived in close proximity to national/republican murals could also be victims of government enforcement or loyalist paramilitaries. Thus, expressions of nationalist/republican culture were more frequently expressed in private: in residences, businesses, sports facilities, Irish dance or language schools, and other "Irish" cultural centers. Even if republicans and corresponding paramilitary organizations attacked the pillars of Protestant rule, public areas, streets, and political murals were perceived as Protestant territory and practice in Northern Ireland. It was not until the 1981 hunger strike that the Catholic/nationalist/republican community began to prominently display murals that supported the republican Hunger Strikers and challenge unionist/loyalist/Protestant-framed narratives of events and belonging.

The emotional and political significance of these murals intensified within these enclaved communities during the 1994 paramilitary ceasefire. The tone and theme(s) of these murals changed as a result of the indeterminate political atmosphere at the time, thereby providing a lens into these communities' perceptions, fears, and reactions to the unfolding political events. For example, concern over the potential for unionists' loss of control of the regional government was expressed in new murals declaring "No Surrender!"

Even after the 1998 peace agreement, which encouraged some to move beyond exclusive binary ethnonational identities, research suggests that this did not occur in "hardened" sectarian areas. Instead, these individuals self-proclaimed a "strong sense of attachment" to one of the binary national identities and felt threatened by the attrition of these previously "set" categories (e.g., Tonge and Gomez 2015). For example, loyalist murals often questioned the state's role in the peace process or demanded the release of paramilitary prisoners during peace negotiations. There also has been, since 1998, an increase and reactive intensity in political murals (i.e., militaristic and combative images), parading/marching, and display of flags in these communities. Many of these murals depicted loyalist paramilitary figures wearing balaclavas; the faceless hostile images appear threatening and intimidating (see Figure 6.2). Similarly threatening militant and aggressive murals of masked republican paramilitaries appeared in Catholic/republican enclaves as well.

Because the majority of these propagandist murals are sectarian, the official policy of the Northern Ireland Housing Executive (NIHE) was to discourage the presence of military or prejudice content in murals on residences. However, de facto enforcement and policing of mural content is managed by local residents and neighborhood leadership.

It is important to note that the messages and images included in these emotional works are highly subjective and therefore are not universally supported by all community members. For example, not all within the local communities appreciated the aggressive militaristic nature of these new murals. Some expressed safety concerns and fear of retaliation if a paramilitary mural appeared in the vicinity of their

FIGURE 6.2 UVF mural, Belfast.

Source: Photo by author.

residences. Over time, many enclaves worked to remove many of the most violent images from corresponding murals, with the remaining militaristic murals continuing to be targets of discussion and negotiations (Jarman 2020).

A closer look at this symbolic medium

As previously described, political murals are a visual medium through which power and vernacular territorial claims are communicated at large, particularly in segregated neighborhoods. Easily accessible to the public and local residents, these emblematic sites serve as a vehicle of propaganda to poignantly dictate a specific interpretation. In this way, they have the potential to shape or encourage a particular understanding of an event, location, or ethnonational group. Their messages are often a commentary on political and social conditions. Others are mnemonics of the past, designed to evocatively foster a shared sense of collective memory, identity, and heritage in contested space in Northern Ireland.

These murals are also painful reminders of violent loss (e.g., reverence for sacrifices made for their local community) and vehicles through which responsibility for the loss is assigned. These are particularly effective if the mural is located close to where a tragic or revolutionary event occurred, transforming the place into a site of memory. There are numerous examples of highly evocative symbolic murals depicting battles or tragedies at the event's location that become revered markers of territorialized commemoration. Some of these murals include painted representations of the faces of murdered local paramilitaries, civilians, and children.

At the same time, these murals reinforce suspicion or hatred of the "Other," or challenge rival enclaves' right to belong (McDowell 2008a). Indeed, critics of sectarian murals regard the impact these images have on local children and their ability to propagate generations of hatred or mistrust. As one local inhabitant of the Shankill neighborhood (P) explained:

> I've lived by a paramilitary mural since the Troubles. Young children walk by murals of masked gunman here and I think, it has to influence them. Doesn't it teach them violence and fear? These murals are asking these children to remember atrocities of the Troubles, but those events that happened before they were born. These murals don't allow children move forward, move beyond the past here.
>
> *(personal interview in Belfast, June 2018)*

Muralist Mark Ervine, son of a former UVF member, expressed similar sentiments in the 2012 documentary "*Art of Conflict: The murals of Northern Ireland*" when he explained:

> It's not hard to pollute an innocent mind. I believe we should be careful about what we show our children, because we're the people that's gonna shape them. And we need to be fearful that we don't shape monsters.

Many political murals venerate and glorify paramilitary violence, commemorate the fallen, and mark territory through messages and symbols of aggressive hatred and bigotry. These are intended to intimate "outsiders" and encourage local residents not to challenge the murals' "validity." Indeed, particularly during periods of acute political turmoil during the Troubles, many murals depicted radical messages, tried to garner support for local paramilitary organizations, and foster mistrust of the "Other."

It is also significant to recognize that most muralists, those who commission the works, and individuals depicted in murals are predominantly adult men. This suggests a limited and gender-biased portrayal of events and commemoration. It also implies the selective exclusion of non-adult male individuals from these works, prompting some to criticize the murals for only presenting a prejudicial view that omits experiences, contributions, and illustrations of women, non-binary individuals, children, and the elderly (McDowell 2008a). Indeed, despite the subjectivity included in any form of art, many interpret murals as "evidence" or a "simple truth" of the past or depictions of rival enclaves. As collective human memory is predisposed toward the visual, images included in the murals attempt to foster a communal "remembering." In this way, collective memory is a device that helps aid the formation of communal bonds and collective forgetting.

Over time, murals have been created, modified, removed and, in some cases, strategically replaced. Since murals can be altered, updated, and adapted to respond to contemporary events, they also function as a barometer for communication. More specifically, murals can serve as a window into perceptions, memories, and memorialization in a local community. They can share messages of peace and encouragement for reconciliation. Murals can also bear witness to a historical or recent trauma as community members grapple with living under the weight of Northern Ireland's past. Indeed, many have become locally revered sites, often serving as a political backdrop for promotional photos and public interviews.

Examples of unionist and loyalist murals

With a long tradition, mural artistry was a constituent form of symbolic assertion of belonging for much of Protestant society before the establishment of Northern Ireland. Unionist or loyalist murals, which are vernacularly identified as "Protestant", are frequently historical in nature, depicting figures and events such as King Billy at the Battle of the Boyne on "the Twelfth" of July 1690,[7] to highlight the community's contributions to the British state and sacrifices to maintain their territorial hold in Ireland.

The first recorded unionist/loyalist mural commemorates King William III's victory at the Battle of the Boyne (see Figure 6.3). Other historical military battles commonly memorialized in murals include the Siege of Derry (1688–1689) or the Battle of the Somme (1916). The Siege of Derry was one of the first major political events of the Williamite-Jacobite War. During the Siege, Protestant residents of Derry's walled city withstood a prolonged attack by Catholic Jacobite forces who demanded the surrender of the city. Refusing to comply and wait for reinforcements, a small group of young men defiantly shouted, "No Surrender!" Their slogan remains a rallying call for many unionists/loyalists as a reminder of their commitment to defend their place in Ulster.

Similarly, the WWI Battle of the Somme, during which the 36th Ulster Division suffered immense losses, is proudly depicted as definitive proof of their obstinate British identity and devotion to the crown (e.g., Graham and Shirlow 2002). Murals featuring the queen or other British political figures are also displayed throughout sectarian loyalist neighborhoods in Northern Ireland to demonstrate their loyalty to the UK. Many of these murals include images of shield crests and political flags (see Figure 6.4) that endow them with great historical and political significance underpinning their importance within their community. Commemorating past events through public murals asserts a sense of belonging and an

7 Despite their importance within the community, some of the historic-themed murals such as King Billy are recently showing signs of aging in certain urban areas. This may suggest a recent lack of emphasis on the maintenance of this symbol.

FIGURE 6.3 An example of a King William III mural, Belfast.

Source: Photo by author.

FIGURE 6.4 Red Hand of Ulster, Shankill mural, Belfast.

Source: Photo by author.

obligation to safeguard their territory throughout the generations. For example, the Red Hand of Ulster, a central symbol in Ulster's provincial flag, is an ancient emblem employed by both republicans and loyalists in Northern Ireland. But for loyalists, the hand is incorporated into paramilitary crests (e.g., UVF, UFF) and insignias for organizations such as the LPA (Loyalist Prisoners' Association), among others. The hand is also depicted outstretched in welcome or clenched into a fist to demonstrate power and devotion to loyalist neighborhoods like Shankill in Belfast.

During socio-political moments of insecurity, unionists/loyalist murals became more militaristic, often demonstrating support for paramilitary forces. Indeed, declarations of fidelity to paramilitaries and the UK were professed more intensely by those who believe their ethnonational identity, community, and way of life are under threat within Northern Ireland (e.g., Nolan and Bryan 2016). For example, when peace talks included the Republic or Sinn Féin in discussions, new loyalist murals emphasized British symbols (e.g., Union flag, Ulster Flag, and the flag of Scotland) and the Red Hand of Ulster in reaction to the unfolding social change in Northern Ireland. Many of corresponding murals were militaristic, demonstrating their indignation at evolving political events (e.g., the AIA) (see Figure 6.5). Particularly for loyalists, the erosion of the state's sole involvement in Northern Ireland provoked many visual efforts to fortify their territorial claim in Ulster.

While murals are a medium through which some members of a communal enclave narrate history or respond to a recent event, not all local inhabitants share the same interpretation. For example, many of the loyalists' overly militaristic murals became targets of contestation, even within the ethnonational enclaves in which they were painted. As a result, some locals organized to remove or replace particularly sectarian murals with messages of reconciliation and cross-community building (see Figure 6.6).

FIGURE 6.5 Loyalist paramilitary mural, Belfast – circa 2014.

Source: Photo by author.

FIGURE 6.6 Replacement mural: Lower Shankill Angels mural, Belfast.

Source: Photo by author.

Examples of nationalists and republican murals

In comparison to themes commonly included in unionist and loyalist murals, nationalist and republican murals, commonly categorized as "Catholic" in the vernacular, often emphasize more recent events and international themes. The 1981 hunger strike was a seminal moment for national/republican muralists. During this time, murals became a medium to promote republican messages, which had previously been systemically silenced and blocked from the media (as discussed in Chapter 4). Indeed, murals became an avenue of defiance, particularly to condemn the state censorship of republicans or their images during the Troubles. In order to bypass the ban, murals with images of these political figures' faces that advertised their struggle provided republicans and the media a "loophole" around regulations as the murals appeared in the background of news reports.

Support for the hunger strike developed from simple graffiti for the H-block prisoners during the blanket protest. The work then evolved into more detailed art forms that included flags, the letter H, and/or other symbols signifying their support for the prison protest (see Graham and McDowell 2007). Eventually, designs included full murals of the hunger strikers and their campaign. Indeed, the strike was a watershed movement for these muralists as numerous works depicting their

FIGURE 6.7 Republican mural commemorating the Rising, Belfast.

Source: Photo by author.

protests or support for republican paramilitary organizations proliferated on the republican landscape at this time (see Figure 6.7).

Thus, in contrast to the gradual development of loyalist murals to include contemporary paramilitary images, the tradition of republican mural work began with their inclusion. As stated in Chapter 4, republicans carefully selected the smiling image of Bobby Sands in their propaganda campaign to represent the 1981 hunger strike (see Figure 6.8). The following section discusses other common images in republican murals.

Not all nationalist/republican murals depicted contemporary events of the time. Others displayed Irish myths, historical Irish political figures, and political events that signified Irish heritage and territorial claim to Ireland. Many included words or phrases written in Irish-Gaelic, as the language holds a strong association with Irish heritage – often serving as an identity marker and a form of linguistic defiance against British rule in Ireland.

Through revolutionary art, these murals promoted a counter-narrative that challenged the socio-political status quo of Northern Ireland at that time. Many of their murals coincided with or bolstered republican territorial claims during a period of rapid urban segregation and self-barricading to protecting their enclaves (discussed in Chapter 5). For example, when Londonderry/Derry's republican Bogside neighborhood created a "No-Go" area and declared their geopolitical

FIGURE 6.8 Bobby Sands memorial mural, Northern SF headquarters, Belfast.

Source: Photo by author.

message of resistance with a mural that states: "You are now entering Free Derry." The "Free Derry" mural created a visible territorial marker and focal point for local commemoration and gatherings (see Figure 6.9).

Depictions of republican paramilitary figures were also commonly included in political murals. Others included messages of bipartisan support for community protests against oppression of Catholics, the injustices of Stormont, police and army violence, and Westminster turning a blind eye to abuses of its Catholic citizens in Northern Ireland. More specifically, one of the most important themes included in many republican murals is repression and resistance. Key symbols of resistance to repression include images of bound hands breaking their chains, Irish mottos "Tiocfaidh ár lá" (Our day will come), or "Saoirse" (Freedom), and phoenixes (i.e., rising from the ash). There are also several murals that commemorate the victims of loyalist paramilitary attacks and the British army. This includes the victims of Bloody Sunday and Bombay Street riots (discussed in Chapters 4 and 5). A few depict the children who were killed by the RUC's plastic bullets fired at civilians during the Troubles. In one mural, the victims' likenesses stand in a line, each behind a burial cross with their names.

FIGURE 6.9 Free Derry mural, Londonderry/Derry.

Source: Photo by author.

FIGURE 6.10 Frederick Douglass mural, Belfast.

Source: Photo by author.

Finally, republican murals also include support and solidarity for international social justice campaigns that highlight discrimination perceived as similar to their own. This has included international anti-imperialist struggles in the Spanish Basque Country or Cataluña, Namibia, Cuba, or historical anti-slavery campaigns such as African Americans' struggle for freedom (see Figure 6.10).

The changing landscape

Over time, political murals in Northern Ireland have become internationally recognized. These eye-catching art forms have been described as "visual media sound bites" of political opinions regarding heritage, identity, and territoriality (Orengo and Robinson 2008). Due to their increasing international notoriety, bourgeoning enterprises within key cities, such as Belfast, began capitalizing on "conflict tourism" to offer tours of these colorful, commemorative, and provocative murals. Perhaps paradoxically, the sites and symbols of conflict in Northern Ireland have become a significant part of the region's tourism.

While "conflict" or "heritage" tourism could be considered a sustainable form of tourism, manipulating or commodifying these landscapes for tourist consumption is problematic. As McDowell, a political geographer in Northern Ireland, argues, the commodification of these murals has serious socio-political, ethnonational, and economic ramifications (McDowell 2008b). For example, "Black Cab tours" in Belfast provide clients private tours of several murals and peacelines as well as the opportunity to talk with locals of the most segregated enclaves in the city. Many such businesses now advertise online, in popular tourist guides, and at local airports and train stations. However, the narration of historical events and the proffered significance of the murals, as well as which murals the clients visit, are influenced by the drivers' ethnonational preferences and personal biases.

Keenly aware of the new scrutiny with which many of these murals are judged, enclaves have become more strategic about the presentation of their murals. Cognizant that this powerful symbolic medium can promote a community's particular interpretation of a historical event, bolster a territorial claim, or incite conflict, certain "showcase" murals are relocated to high-visibility locations. In contrast, many of the more aggressive and controversial ones have been removed from public view.

The removal of some of the more offensive murals reflects negotiations between local leaders, muralists, and/or political officials. As the joint task force on reconciliation in Northern Ireland argues, sectarian murals can be devastating to peace efforts because they are:

> visible legacies of our conflict, glorifying, even sanctifying paramilitaries, and the "romance" of the gun; commemoration sites saluting respective "war" dead as heroes. Similarly, painted kerb-stones marking tribal turf; confrontational flag-waving; and other symbols of aggressive partisanship, bigotry and hatred, scar the landscape of many towns and cities. In effect, demarcating exclusive ethnic terrain that are hostile and "No-Go" areas to all outsiders, they can be intensely intimidating. Highly questionable is the extent to which these depictions are truly the violence and choice of the local people as distinct from the stance of militia organisations.
>
> *(Gaffikin et al. 2016)*

In 2009, the regional government launched a "re-imaging" program that provides local communities and residents funding to replace offensive, aggressive, and sectarian murals with peaceful art and messages (Community Relations Council 2009). More recently, politicians vowed to remove all sectarian murals by 2023.

The art of reconciliation

Efforts to bridge the chasms between divided ethnonational groups are underway, but progress is slow. Parading continues to be a controversial topic in Northern Ireland, and many sectarian murals remain part of the socio-political landscape. However, there is a growing emphasis on creating murals that are designed to help a community heal from conflict and loss. For example, muralists such as Robert Ballagh and the artists' cooperative "Art of Reconciliation" in Derry/Londonderry are internationally recognized for their promotion of reconciliation, peace, and healing through their artwork.

The political and artistic practices of murals continue to be viewed with new interpretations as their audience, the murals, and their political underpinning change. Murals, engendered with great political significance, produce a kaleidoscope of meanings for residents of an enclave and those outside of it. Contestation over murals remains a key issue within Northern Ireland as disparate perspectives reveal highly divided political realities and interpretations of the past and present. The implications and contestation stretch across local, regional, national, and international scales, as both communities share a legacy of longing and loss, spatially reified on the local landscape.

The following chapter explores how the local landscape is employed in cross-community building and reconciliation efforts. More specifically, Chapter 7 examines a "shared" cross-community center in Belfast that works to forge meaningful relationships across sectarian divisions.

References

Bardon, Jonathan. 1992. *History of Ulster*. Belfast: Blackstaff.

BBC. 2017. "Height of Bonfires Limited by Government." July 11, 2017. www.bbc.com/news/uk-northern-ireland-40566046.

Cohen, Shaul. 2007. "Winning While Losing: The Apprentice Boys of Derry Walk Their Beat." *Political Geography* 26 (8): 951–67.

Community Relations Council. 2009. *Towards Sustainable Security: Interface Barriers and the Legacy of Segregation in Belfast*. Belfast: Community Relations Council. www.community-relations.org.uk/.

Corrymeela. 2018. "Corrymeela Community." www.corrymeela.org/.

Dempsey, Kara E. 2012. "'Galicia's Hurricane': Actor Networks and Iconic Constructions." *Geographical Review* 102 (1): 93–110.

Gaffikin, Frank, Chris Karelse, Mike Morrissey, Clare Mulholland, and Ken Sterrett. 2016. *Making Space for Each Other: Civic Place-Making in a Divided Society*. Belfast: Queen's University Belfast.

Graham, Brian, and Sara McDowell. 2007. "Meaning in the Maze: The Heritage of Long Kesh." *Cultural Geographies* 14 (3): 343–68.

Graham, Brian, and Peter Shirlow. 2002. "The Battle of the Somme in Ulster Memory and Identity." *Political Geography* 21 (7): 881–904.

Hagen, Joshua, and Robert Ostergren. 2006. "Spectacle, Architecture and Place at the Nuremberg Party Rallies: Projecting a Nazi Vision of Past, Present and Future." *Cultural Geographies* 13 (2): 157–81.

Jarman, Neil. 2020. *Material Conflicts: Parades and Visual Displays in Northern Ireland*. Oxfordshire: Routledge.

Johnson, Nuala C. 2002. "Mapping Monuments: The Shaping of Public Space and Cultural Identities." *Visual Communication* 1 (3): 293–98.

McAuley, Jim. 2016. "Memory and Belonging in Ulster Loyalist Identity." *Irish Political Studies* 31 (1): 122–38.

McDowell, Sara. 2008a. "Commemorating Dead 'Men': Gendering the Past and Present in Post-Conflict Northern Ireland." *Gender, Place & Culture* 15 (4): 335–54. https://doi.org/10.1080/09663690802155546.

———. 2008b. "Selling Conflict Heritage through Tourism in Peacetime Northern Ireland: Transforming Conflict or Exacerbating Difference?" *International Journal of Heritage Studies* 14 (5): 405–21.

McDowell, Sara, Maire Braniff, and Joanne Murphy. 2015. "Spacing Commemorative-Related Violence in Northern Ireland: Assessing the Implications for a Society in Transition." *Space and Polity* 19 (3): 231–43.

Nolan, Paul, and Dominic Bryan. 2016. *Flags: Towards a New Understanding*. Belfast: Institute of Irish Studies, Queens University.

Nolan, Paul, Dominic Bryan, Clare Dwyer, Katy Hayward, Katy Radford, and Peter Shirlow. 2014. *The Flag Dispute: Anatomy of a Protest*. Belfast: Queen's University Belfast.

Orengo, Hector A., and David W. Robinson. 2008. "Contemporary Engagements within Corridors of the Past: Temporal Elasticity, Graffiti and the Materiality of St Rock Street, Barcelona." *Journal of Material Culture* 13 (3): 267–86.

Parades Commission. 2020. "Parades Commission for Northern Ireland." www.paradescommission.org/getmedia/73089983-f4b8-46a8-b816-7ea709122d81/NorthernIrelandParadesCommission.aspx.

Press Association. 2018. "Orange Order Parades Take Place amid Violence in Northern Ireland." *The Guardian*, July 12, 2018. www.theguardian.com/uk-news/2018/jul/12/northern-ireland-orange-order-parades-12-july-amid-violence-belfast-derry.

Rolston, Bill. 2012. "Re-Imaging: Mural Painting and the State in Northern Ireland." *International Journal of Cultural Studies* 15 (5): 447–66.

Tonge, Jonathan, and Raul Gomez. 2015. "Shared Identity and the End of Conflict? How Far Has a Common Sense of 'Northern Irishness' Replaced British or Irish Allegiances since the 1998 Good Friday Agreement?" *Irish Political Studies* 30 (2): 276–98.

7

HOPE FOR THE FUTURE I

"Shared" space in north Belfast

The previous two chapters explored exclusionary national rhetoric, segregation, and practices such as parading, flagging, and political murals that function as powerful spatial manifestations of divided ethnonationalism. After the 1998 peace agreement, Northern Ireland continues to face formidable challenges as a post-conflict society, including combating the vestiges of sectarianism and segregation. As discussed in Chapter 5, the violence and intimidation that became so prevalent in certain parts of Belfast during this conflict profoundly shaped the city's spatial distribution. This resulted in a concentrated patchwork of fractured areas of deeply divided ethnic enclaves that closely border one another (e.g., Gaffikin et al. 2016). For many inhabitants of these divided neighborhoods, borders do not terminate at the end of peacelines. Instead, many residents developed mental maps underpinned by perceived areas of difference, danger, and safety that cognitively shaped individuals' movement throughout the city.

While geographic segregation and isolation can contribute to a sense of cohesion within each of the enclaved communities, it also perpetuates animosity and mistrust of the "opposing" community. Without opportunities for positive interaction, many children born after the Troubles learned to fear, and even hate, members of a "rival" ethnonational community without meeting or interacting with them. Furthermore, research suggests that transmission of a previous generation's memory could occur if the experiences communicated are powerful and traumatic, thereby impacting members of the subsequent generation "so deeply as to constitute memories in their own right" (Hirsch 2008, 103).

Despite the turmoil, many individuals and "cross-community" organizations combat exclusionary, antagonistic, and essentialized expressions of sectarianism. This includes individuals who live in segregated enclaves, as one co-director of

DOI: 10.4324/9781003141167-7

a cross-community organization in Belfast explained the origins of his interest in peace work:

> When I was young, a small boy rang our door at night. He was crying. He had been chased by older boys and now was lost. My ma asked where he lived. We were surprised to learn he was from the neighboring Catholic area. The boy didn't know he had just knocked on the door of a family of [loyalist] UDAs! Now, my da always said – "never trust Catholics, our enemy," so, I thought my ma would turn the boy away. Instead, she told my brothers to walk the boy back to home. I remember her saying, "boys, get this young one home safe." And my brothers did. It was the first chance to see a Catholic as a *person*. I hadn't met one before and here was my ma making sure he got home safe. It was a profound moment for me, even if I didn't realize at the time. Looking back, I think her actions demonstrated universal human kindness. And, I've been building on that through my work here at our local cross-community center.
>
> *(personal interview in Belfast, 2018)*

Indeed, grassroots organizations in contested areas throughout Northern Ireland are breaking down barriers to combat sectarianism. To ameliorate the ethnonational segregation that scars much of the region, organizations and individuals are creating "shared space" where people can safely cooperate on peacebuilding efforts and dismantle the spatial politics of sectarianism.

This chapter continues the examination of borders, landscape, conflict, and reconciliation through an investigation of the role of local grassroots "shared space" in segregated Belfast. Through a case study, this chapter explores how these spaces are created and utilized to combat the territorial consequences of enclaves and exclusionary forms of ethnonationalism. More specifically, this chapter examines the efforts of 174 Trust[1] (hereafter the Trust), a grassroots cross-community "shared center" in the New Lodge (C) neighborhood of northwest Belfast. The Trust's mission was to foster a safe "shared place" where local residents from neighboring sectarian enclaves can gather for cross-community programming to combat sectarianism (174 Trust Mission Statement 2018). Through institutional adult and child programs, daycare, and sports leagues designed with peacebuilding, respect, and collaboration at their core, the Trust facilitates much-needed interaction, dialogue, and cross-border relationships in north Belfast.

The next section discusses how geographers understand the employment of landscapes, particularly in peace efforts and as "shared space." This is followed by a brief history of sectarianism in the New Lodge neighborhood and the mission of the Trust. This chapter then examines the implications of many of the programs offered to local

1 This chapter draws from in-depth, longitudinal interviews conducted over 4 years with the director, employees, people who use this center, local community members, municipal and national politicians, as well as archival research and participant and landscape observation (Bryman 2016).

residents at the Trust, including some of its notable reconciliation-related achievements within the community. Finally, this chapter concludes with a reflection on the lessons learned from this local grassroots cross-community reconciliation work.

Addressing a legacy of violence and spatial division in a post-conflict society

Despite the close physical proximity of many of Belfast's sectarian-divided enclaves, some Northern Irish natives who have subsequently moved to the city observed how intensely segregated it remains after the Troubles. As one employee of the Trust explained:

> When I first moved to Belfast, I was astonished by how habitually sectarian narratives are drummed into individuals in north Belfast by friends, families, and fellow community members. In this part of the city, the divided communities have segregated themselves – they attend separate and segregated schools, churches, clubs and shops, which further reinforces the divisions here.
>
> *(personal interview in Belfast, 2016)*

Indeed, powerful sectarian narratives and fear continue to galvanize the divisions that exist between these entrenched communities (e.g., personal interviews conducted in Derry, Coleraine, Belfast 2015–2018; McAuley and Ferguson 2016). The historical and contemporary presence of paramilitary forces in these communities also reinforces sectarianism (e.g., Mesev, Shirlow, and Downs 2009). For example, some actively circulate narratives detailing the imminent threat posed by residents in "opposing" enclaves and enforce their own self-policing and retaliation policy should their local residents associate with members of a rival enclave (e.g., Dowler and Ranjbar 2018; Cunningham and Gregory 2014).

In the unstable atmosphere of the highly divided sectarian areas of Belfast, it is rare to encounter places that are jointly utilized or considered safe by both communities. Indeed, residential segregation remains particularly problematic in places such as Belfast, especially in north, west, and eastern Belfast, where sectarianism is profound (e.g., Nolan 2017).

Since the 1998 peace agreement, local, regional, state, and EU governments have provided millions of Pounds and Euros in the spirit of the "Contact Hypothesis," as discussed in Chapter 5. These efforts focus on creating and supporting programs that provide residents opportunities for positive interactions with members of neighboring enclaves (e.g., McDowell, Braniff, and Murphy 2017). As evidence suggests, cross-community interactions instituted in trust can encourage individuals from different ethnonational backgrounds to build intercommunity relations that may not have been otherwise forged. Thus, the aim of these endeavors underpins the hope to foster positive cross-community dialogues to reduce prejudice, xenophobia, and sectarianism in Northern Ireland (Lloyd and Robinson 2011). The belief in the essentiality of forging cross-community relations across perceived and physical borders of division is shared by many throughout Ireland. It is also at the heart of the satirical Twitter account @BorderIrish, as the "Irish border" explains:

> The most important thing I've learnt is this: borders [physical and perceived] are the most cowardly form of human interaction. Opening yourself up to strangers, opening yourself up to the new and the unknown and the unexpected – that's bravery.
>
> *(2019, 249)*

The significance of (shared) space

Geographers have long theorized about the role of space in conflict, territorialization of violence, and assertions of spatial control. For much of Northern Ireland, the socio-spatial context of violence and segregation patterns along sectarian lines are catalysts and manifestations of these forces. At its core, the conflict in Northern Ireland underscores struggles over sovereignty and power, territory (i.e., the desire to claim, control, and belong), and sectarian ethnonational identities (McDowell, Braniff, and Murphy 2017; Graham and Nash 2006). Indeed, much of the geography of exclusion in Northern Ireland could be described as "spaces of hate"; as individuals and groups compete to control space, they reinforce ethnonational territoriality along the lines of bounded delineation and marginalization (Flint 2004).

However, space can also be utilized to combat sectarian divisions that exist in everyday space. Despite a legacy of conflict and resistance to change, "shared" space is a crucial and transformative element for fostering engagement and integration in divided communities (e.g., McDowell, Braniff, and Murphy 2017). The creation of a shared place dedicated to reconciliation in a post-conflict society can foster tolerance and bridge divisions within. This will not happen spontaneously; rather, it must be deliberately facilitated, designed, and managed at the local level for integration and peacebuilding (Gaffikin et al. 2016).

To challenge the reproduction of segregated geographies in north Belfast, organizations such as the Trust provide a shared place to meet, engage, and actively work for reconciliation. A shared place must be safe and adaptable to provide a variety of opportunities for peacebuilding efforts (McDowell, Braniff, and Murphy 2017). Indeed, to overcome the territorial consequences of sectarianism, people need the opportunity to gather face-to-face, share stories, "learn" peace, plan for the future, and discuss the past (174 Trust Newsletter 2018; Corrymeela 2018). Without shared places, peacebuilding efforts often remain foundationless and ungrounded.

The shared future initiative

In 2003, the Community Relations Unit in Northern Ireland sponsored the Harbison Review's *Shared Future: Improving Relations in Northern Ireland* initiative. This lengthy consultation process featured community dialogues to develop a set of policies that promote principles "for a plural but socially cohesive society and a series of policy options for fostering 'good relations' in Northern Ireland" (Graham and Nash 2006, 253). As a result of some failed projects, there was some skepticism regarding the efficacy of shared space. However, most unsuccessful attempts were state-imposed "top-down governance" that failed to engage or collaborate with local communities (Gaffikin et al. 2016; McDowell, Braniff, and Murphy 2017). As the following sections demonstrate, local community-based organizations have been more successful. With support, several grassroots organizations developed cross-community programming predicated on the creation and utilization of shared space. For example, the Junction in Londonderry/Derry provides a "space for activities that feed into the development of relationships, better understanding and mutual respect" to support cross-community dialogues and interactions between various individuals, including former paramilitary combatants (*http://thejunction-ni.org*).

Similarly, the *Forgiveness Education Programme*'s special curriculum is incorporated into daily academic lesson plans in participating "shared" or "mixed" schools and is dedicated to cross-community education and relationships. These lessons, tailored for students in primary years 2–7, are designed to help develop a greater understanding of reconciliation and inherent worth of all individuals, regardless of creed, ethnicity, race, age, or gender, particularly through an emphasis on empathy and positive interactions with a member from an "opposing" community (Enright 2013).

Comprehending the impact that such places can have on divided societies, the future director of 174 Trust set out to create an intentionally shared place in the profoundly segregated neighborhoods that comprise northwest Belfast. Once established, the Trust became a non-denominational community and performance center in the New Lodge (C) neighborhood, an area with strong republican ties, and directed by Bill Shaw – a local who grew up in one of Belfast's loyalist neighborhoods. The mission of this social justice organization is to provide a safe and inclusive place for all who enter, regardless of religion, gender, age, or sexual orientation (174 Trust Newsletter 2018). The Trust serves as a bridge for residents of divided enclaves to elude physical and perceived sectarian borders. Its focus on cross-communal inclusion, peaceful dialogue, respect, and reconciliation is the core of its success. Since the Trust has become a beacon of hope for beleaguered north Belfast, director Shaw received the Order of the British Empire (OBE) (e.g., www.globalpeace.org/media-gallery/detail/4058/8204). This accommodation, appointed by the British crown, honors his work promoting peace and reconciliation through a shared space in an otherwise deeply divided area.

The New Lodge: a sectarian neighborhood in north Belfast

While certain areas of Belfast, such as the riverside redevelopment along the River Lagan, have benefited from urban renewal and gentrification efforts, north Belfast remains one of the most impoverished, with "some of the highest levels of poverty and social deprivation in Western Europe" (Campbell 2017). While the mean mortality rate for Northern Ireland is 80, north Belfast's average is considerably lower at 66. In addition, in some of the neighborhoods in north Belfast, 53% of inhabitants have achieved the lowest educational qualifications (174 Trust 2018). However, instead of efforts to ameliorate this deficit, the municipal government effectively overlooked this area's infrastructural or development needs for the last 40 years (Gaffikin et al. 2016).

The Trust is located within this disadvantaged part of the city on the Antrim Road in the New Lodge (C) neighborhood. In the nineteenth century, the New Lodge was once a predominantly Protestant area as a result of the proliferation of industrial jobs that developed in Belfast when the city became one of the industrial centers of the British Empire. However, when this intercity neighborhood suffered infrastructural damage in 1941 during the Belfast Blitz, the destruction sustained during the war contributed to a process of suburbanization for Protestants with greater financial means. In addition, when the golden age of working-class employment weakened as a result of outsourcing and deindustrialization, joblessness became progressively widespread for many of the remaining inhabitants of the New Lodge.

Gradually, the ethnic composition of the New Lodge became increasingly mixed, until sectarian violence during the Troubles produced a sharp decline of Protestant households residing in the area. Neighborhood exodus to ethnic enclave "sanctuaries" was common for any ethnonational minorities in contentious neighborhoods in Belfast. Indeed, residential segregation along sectarian lines, especially in the economically weak area of north Belfast, became more delineated as a result of violence and intimidation. As discussed in Chapter 5, the city became a patchwork of fractured sectarian enclaves and peacelines. During the height of the violence during the Troubles, the New Lodge (C) became a Catholic/republican enclave bordered by Protestant/loyalist Shankill (P) and Crumlin (P) enclaves to the west and Duncairn (P) to the north and east. As a result of its bounded geographic location and republican affiliations, the New Lodge became a target of loyalist violence. Subsequently, the periphery of this neighborhood, the Antrim Road, was known as "Murder Mile" due to the high number of its inhabitants who were killed by loyalist paramilitaries (see Map 7.1).

As violence during the Troubles persisted, community divisions deepened as many of these geographically segregated communities became more volatile and radical in affiliation. This included the New Lodge, where there was a notably high amount of paramilitary activity (i.e., IRA, PIRA, and the UDA, UVF) throughout the neighborhood. Fractious sectarian parades in this area also provoked and renewed tensions between the communities, often igniting riots when marches traversed interface boundaries. During the Troubles, gates along the local peacelines closed nightly to prevent kidnapping or murder gangs from entering enclaves, and the contiguous communities became increasingly sequestered and embittered. However, the fear of continued or rekindled sectarian violence persists in problematically segregated areas such as the New Lodge, where gates in the peaceline continue to be closed periodically in 2021.

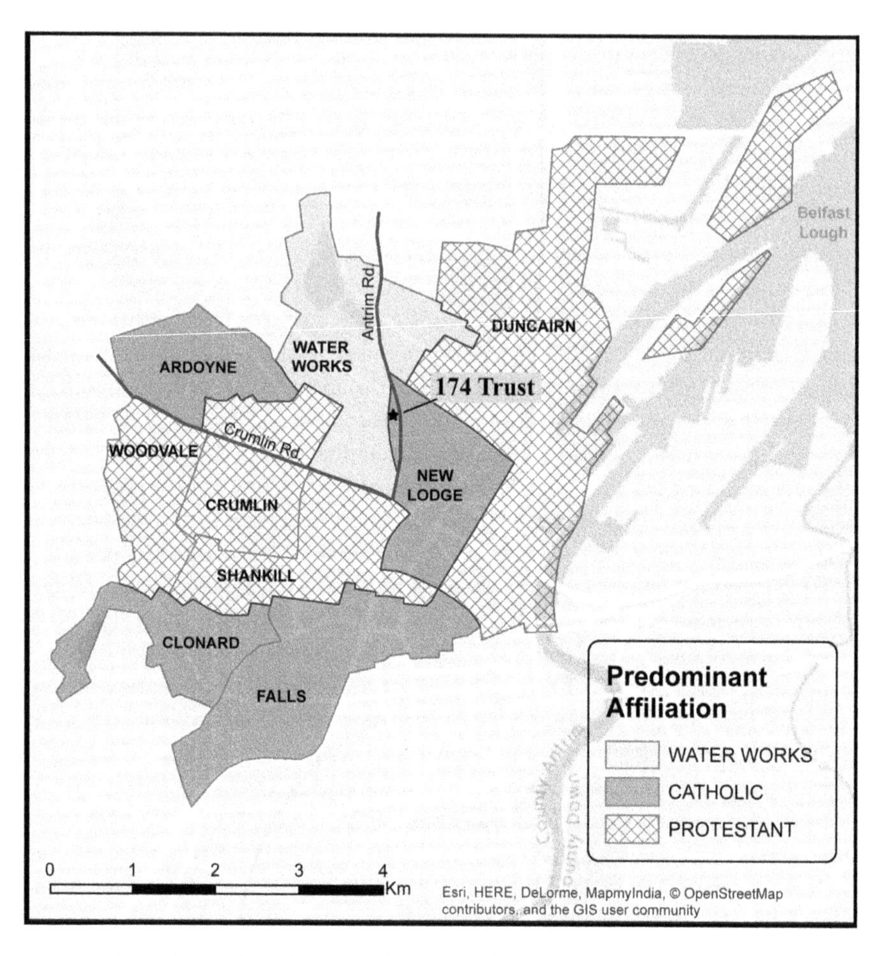

MAP 7.1 Location of 174 Trust and its surrounding neighborhoods

174 Trust

Despite deep and protracted fault lines, not everyone in these enclaves believes that sectarian divisions are intractable. For example, the director of the Trust, Bill Shaw, is a firm proponent of integration and considered the location of the Trust (on the Antrim Road in the New Lodge) as an auspicious geographic location to create a "shared" community center in north Belfast. Since the New Lodge adjacently borders Protestant/loyalist Shankill, Crumlin, and Duncairn, he wanted to capitalize on the geographic proximity of the segregated enclaves. Shaw believed that in order for the Trust to effectively bridge divided ethnonational divisions, participants needed a safe shared space to work together for a common goal. Indeed, he recognized that this part of Belfast had a variety of needs, including opportunities for meaningful cross-communal engagement.

In order to best serve the local communities, he met with local residents to inquire about their needs and foster local proprietorship of the Trust's programs. Shaw's reputation for his mediation work during some of the most contentious moments in the Troubles, such as the Holy Cross dispute (discussed in Chapter 5), earned him the respect of many New Lodge inhabitants. His standing as a leader who was respected on both sides of the sectarian divide is particularly significant, especially as the need for strong cross-community leadership is one of the key recommendations of the Harbison Review for building a *Shared Future* in Northern Ireland.

As a Sinn Féin representative stated in a personal interview at the Trust:

> The transformational power of the Trust is its director, Bill. He is a Protestant who worked for years in Belfast on conflict resolution and reconciliation efforts during the Troubles. Now he runs a community center that helps empower residents of north Belfast. He invites people into the Trust and helps them meet people from other neighborhoods – including youth. That's rare for Northern Ireland and especially rare in north Belfast.
>
> *(personal interview at 174 Trust, 2016)*

Shaw also ensured that the Trust hired employees from both ethnonational backgrounds to have a "mixed" and representative staff. As one employee explained in an interview:

> I previously worked on various short-term cross-community reconciliation projects in Belfast and knew Bill by reputation. When he approached me to ask if I would work for the Trust, he asked me to help him create a shared space so locals could explore their common ground instead of their sectarian differences. But I told him – "Bill, You're a Protestant and originally from loyalist neighborhood, right? You know I'm a Gael and a passionate Irishman, right? But I also told him, Bill, I also love north Belfast, I love this community, the city, and I want what's best for it." And he said, "I know. That's why I'm asking you to help me make this cross-community project work."
>
> *(personal interview at 174 Trust, 2016)*

The *Shared Future* report also identified the need for widespread engagement and ownership, the importance of local action, and targeted assistance for youth and areas with "a high incidence or history of poor relations and conflict and deprivation" (Harbison 2002). These highlighted points are specifically addressed through the programming provided by the Trust. For example, when locals asked Shaw if the Trust could provide space for neighborhood programs such as Alcoholics Anonymous or a support group for parents of children with disabilities, he provided the space, as long as the groups' membership included members of both ethnonational communities (Dempsey 2022).

By providing shared space and program support, the Trust addresses the needs of local individuals. It also empowers local participants by granting them the direction and proprietorship of many of the programs housed within the Trust. These elements are significant. As Contact Hypothesis purports, cross-community relationships are most effectively fostered when disparate individuals have a common goal toward which they can work together and they have the space and proprietorship of organizations to help them achieve their objectives (e.g., Dowler and Ranjbar 2018). For example, Alcoholics Anonymous or the support group for parents with children with disabilities shift the focus away from the sectarian nature of an individual to allow participants to mutually work together to battle a common illness or support the various needs of their children.

The Trust is also an advocate for fair public housing in north Belfast and provides support for community members who feel excluded from municipal programs. This includes individuals with disabilities, mental health and addiction illnesses, and former paramilitary combatants. Its staff, comprising both Catholics and Protestants (as well as members of other religions' traditions and/or agnostics), continue to work with locals to create new outreach for their programs. For example, the Trust developed reconciliation programs that partner with other "shared" spaces programs throughout Northern Ireland. In an interview, the director explained the importance of these networked collaborations:

> The Trust is a shared, safe environment where Protestants and Catholics can meet and engage with each other, where prejudice and sectarianism can be challenged and stereotypes can be shattered. In a segregated society like ours, it's so all too common for people to have no meaningful contact with the other side. But after all my work with the Trust, I am convinced once you encounter someone from the other side in a supportive environment, you will never be the same. You can learn to respect the dignity in others.
>
> *(personal interview at 174 Trust, 2017)*

Youth and Adult programs at 174 Trust

Youth programs

The Trust's Youth Programs, such as its daycare, kindergarten, and several arts and sports-based afterschool programs, are specifically designed to foster positive interactions between "mixed" communities. This includes immigrant and LGBTQ+ individuals, which is uncommon for youth centers in the area (174 Trust 2018). While there is a relatively small percentage of international immigrants in the region, over 1,200 Syrian refugees lived in Northern Ireland by December 2018 (NILT 2021). In September of the previous year, the Trust hosted the Belfast City of Sanctuary Conference that welcomed refugees and asylum seekers in Northern

Ireland. One Syrian parent explained the significance of this commitment in a personal interview:

> We take three buses to bring my daughter to the kindergarten here. We heard that this place [the Trust] is a place where she will be safe, so we come here for her.
>
> *(personal interview at 174 Trust, 2019)*

During the Trust's 2018 kindergarten graduation ceremony, I asked parents/guardians if they selected this program due to its inclusive nature. Many of the parents/guardians responded that it was their primary motive for utilizing the Trust (Dempsey, survey conducted in 2016 and 2018).

Boys' youth "mixed" football team

One of the most instrumental programs is the Trust's Youth "Mixed" Football Team. It assimilates children from the New Lodge (C), Shankill (P), and Crumlin (P) neighborhoods on a single team. Several interfaith organizations such as Corrymeela, Northern Ireland's oldest peace and reconciliation organization, also utilize integrated sports teams as an avenue through which youth can forge positive relationships across ethnonational lines. By forcing a "mixed" group to join together to compete against a "new" opponent (i.e., another team), the ensuing camaraderie helps foster a sense of community for all team members, regardless of ethnonationality (e.g., Bleakney and Darby 2018). As one of the all-Ireland rugby team (IRFU) former captains, Willie John McBride explained in Chapter 1:

> In sports, we don't care about politics . . . we just respect the people you're with on the team.
>
> *(cited in Williams 2018 film)*

The fractured geography of the players' sectarian residences presented a challenge for the team. For example, players living in the Protestant neighborhoods lacked safe walking routes or public transportation to arrive at the Trust. As a result, director Shaw purchased a minibus to drive the teammates to and from practices and games. The minibus also allowed Shaw to take the team into other enclaves and visit other Belfast interfaith "mixed" organizations. These planned excursions provided critical opportunities for what Shaw called "teachable moments" and group dialogues on sectarianism, territoriality, identity, and safety as they traveled throughout the fragmented city over the years.

When Shaw learned that the majority of the teammates had never entered an "opposing" sectarian enclave, he organized minibus trips throughout the segregated neighborhoods. During these trips, he facilitated discussions about identity and ethnonationalism, asking the young players to consider questions such as follows:

> Which of your teammates feels like they belong here? Why?
> Who feels safe here? Do you feel safe here? Why?

> What have people told you about people from "the other side"?
> How does that compare with your actual experiences with your teammates from the "other side"?
> What can you do to help make your mates feel safe and included?
>
> *(personal interview at 174 Trust, Shaw 2018)*

Many of the boys identified symbols such as the presence of paramilitary flags and "peacelines" in addition to stories of sectarian violence as factors that shaped their perception of "opposing" spaces and residents. Shaw then arranged for the team to meet several of his friends' families that live in various segregated enclaves to forge a foundation that fosters cross-community interactions and friendships.

As a result of their supportive experiences for several years, the cross-border friendships the teammates formed were both exceptional within north Belfast and longstanding. Since they frequently played and ate together after practices and games, many maintained their friendships once entering separate high schools. For example, one Shankill (P) teammate invited a teammate from the New Lodge (C) to join his new high school friends at their school dance. According to Shaw, this is an indication of the success of meaningful integration. As one of the teammates explained:

> This year my mate from the Shankill Road invited me to join his mates for their school dance. Religion isn't an issue for us. Other kids in the New Lodge who didn't play on our team don't understand why we're friends. But we went to football matches, traveled, and played together for years. Now that we can drive, it's easier to see each other after school too. I go there, they come here. We all grew up hearing stories about how the "other" guys will hurt you, but when you actually meet someone from the other side, it's different.
>
> *(personal interview at 174 Trust, 2016)*

Adult programs

The Trust also offers adult programs designed to forge positive interactions among residents from sectarian communities (e.g., "mixed" shared-reading program, a women's interfaith group, art workshops, Irish language classes, and support groups for individuals struggling with mental health and/or substance abuse).

The shared-reading program

The Trust's shared-reading program, led by an instructor trained in conflict mediation and literary analysis, facilitates constructive conversations through specifically selected texts. Each literary selection offers members the opportunity to critically analyze themes such as conflict, grieving, loss, forgiveness, and reconciliation. She also invites local authors and poets to speak about their related work with the members and discuss the reading as a group.

In personal interviews, the participants explained that they were initially reluctant to share their opinion about a discussion topic, as many in Northern Ireland are hesitant to publicly speak about sectarianism. However, as members began to share their thoughts with the group, the instructor helped facilitate constructive cross-community dialogues. Interviewees explained that these interactions, which could be contentious at times, provided participants with insight into an "opposing" ethnonational perception about "truths" regarding the Troubles (personal interviews at 174 Trust, 2017).

Eventually, one member invited the entire group to her house for dinner. Others subsequently extended similar invitations. Some attended religious services with members from other denominations. In interviews, participants expressed how they were surprised to learn how similar everyone was regardless of ethnonational or religious background. As one participant explained:

> After all these years believing in the need to stay separated, to maintain the division walls here, we discovered we have more in common than what divides us. Not everyone in the group have become dear friends, but you know, for the first time in my life, I have friends from other religious denominations. I never would have met these people if I didn't join the reading group. It's really profound to think about how, despite our proximity, I would not have made these friends without this reading group.
>
> *(personal interview at 174 Trust, 2016)*

Adult and child programs

The Trust also offers programs designed for both adults and children. For example, there is a group for parents of children with disabilities that provides support for the adults and enrichment activities for the children. During a personal interview with this group's leader, an interview with this group's leader revealed how the group demonstrates the Trust's influence in sectarian north Belfast. The leader's brother was one of the "New Lodge 6" – six young men killed in the Catholic/republican neighborhood in targeted sectarian attacks by the British Army and loyalist paramilitaries in 1973. The program leader explained how the "New Lodge 6" families believe there has been no justice for the unprovoked killings.

While the leader continues to struggle with her familial loss due to sectarian violence, she embraces the "mixed" support group. This is significant as some participants initially displayed loyalist paramilitary flags at their residences. She explained:

> Seeing UDA flags at some of the houses where I drove families in the minibus was agonizing. But we then collectively agreed that our children are the

> priority, so we would set aside sectarian nationalism when we met. It took a long time, but I think our children became the bridge that unites these families.
>
> *(personal interview at 174 Trust, 2016)*

Like the adults in the support group for children with disabilities, adults with children in other programs at the Trust also reported cross-communal friendships forged through their children. For example, after one of the Trust's kindergarten graduation ceremony, students posed with family members, classmates, and friends for pictures. One group included three girls hugging each other, holding a sign that read "BFFs." Parents/guardians explained one girl was a Sudanese Muslim immigrant who moved to Belfast 3 years ago, one was a Catholic from the New Lodge, and the third was a Protestant from Shankill. The families explained that the girls insisted on playing together after school, giving the adults a unique opportunity to develop their own cross-border friendships as well.

Outreach programs

The Trust also hosts international peace conferences and collaborates with cross-community organizations throughout the UK. For example, "Together, Stronger" offers:

> multi-stakeholder programs that are established to discuss, identify and advance social and physical regeneration in the area . . . and aims to boost regeneration and community cohesion in north Belfast.
>
> *(IFI 2018)*

Since partnering with "Together, Stronger," the Trust now offers additional integrated sports teams, community theater, and various skill-based training classes. According to Belfast's Lord Mayor, Councilor Niall O'Donnghaile:

> The 174 Trust has put in place projects that can boost the social fabric of North Belfast and has the potential to become a beacon for other local communities. By increasing dialogue and participation in cross-interface activities, there is an opportunity to build safe, shared spaces and foster future peace building efforts. It is important that communities can come together on shared issues and find solutions that can benefit everyone.
>
> *(174 Trust 2018)*

Other outreach programs include local residents in campaigns to redesign political murals in north Belfast. For example, one group removed a contentious sectarian mural that celebrated Oliver Cromwell's conquest of Ireland from the Shankill neighborhood (see Figure 7.1). They replaced it with a memorial dedicated to peace and solidarity that reads, "Remember, Respect, Resolution" (see Figure 7.2).

FIGURE 7.1 (Former) Oliver Cromwell mural, Belfast.

Source: Photo by author

FIGURE 7.2 Replacement triptych (of Cromwell mural), Belfast.

Source: Photo by author.

This triptych is a physical manifestation of local commitment to erode sectarian legacies in the local environment. While this act may appear trivial to an outside observer, Shaw explained:

> Twenty years ago, you would never have seen anything so inclusive in north Belfast. But many of us here are decided to share a narrative of an inclusive place where we work together to reach our goals. In the end, this is our shared place and Northern Ireland is our home. I know we still have a long way to go before we see widespread tolerance and inclusion in Belfast, and particularly in enclaved places like New Lodge and Shankill, but victories like this mural continue to give us hope for the future.
>
> *(personal interview at 174 Trust, May 2016; cited in Dempsey 2022)*

Prior to the Trust's establishment, north Belfast was the only part of the city that lacked a local art/cultural center. Today, its "Duncairn Centre for Culture and Arts" provides "mixed" academic, art, and music programs for local residents. In addition to its café, art galleries display local art, and art studio space is available to graduate students as Artists in Residence. The Trust's Duncairn Centre also has a soundstage with weekly musical performances designed to bridge sectarian communities. Both Shaw and the musical coordinator agreed that the venue should be a space that belonged to all. Therefore, the coordinator designs the musical schedule to feature performers that appeal to a mixed audience. While some of the performers are local, others are internationally recognized (theduncairn.com). The diverse musical performances have successfully drawn audiences from all parts of Belfast, thereby increasing the Trust's recognition (174 Trust 2018). Some of the more renowned headliners' performances are recorded and broadcast throughout the UK and the Republic by BBC (British Broadcasting Corp.) and RTE (Ireland's National Radio/TV broadcasting).

The Trust's art classes are taught by employees and/or instructors that collaborate with other local community arts organizations in Belfast. This makes art instruction more accessible in north Belfast and supports the dynamic citywide network of art organizations. As the Chief Executive of the Arts Council of Northern Ireland explained:

> Establishing a multi-purpose arts centre of world class in North Belfast has placed arts and culture right at the heart of local community life. It also has provided a safe and welcoming place for people and communities to come together to share in the positive experiences associated with the arts. It is transforming pride and opportunity in the area. It is, in short, changing lives.
>
> *(174 Trust 2018)*

Lessons from a shared place

While critics may suggest the Trust only impacts only a few neighborhoods, this grassroots organization successfully fosters cross-border community involvement in geographically fractured and beleaguered north Belfast. As Northern Ireland continues to evolve as a post-conflict space, multiple interlaced mechanisms such as economic regeneration and shared political governance are necessary for effective societal transformation. However, simultaneously, grassroots organizations are essential for combating engrained geographies of residential divisions (e.g., Kasirova 2014). As the Trust's programming, employees, and participants demonstrate, it has become:

> a transforming force in the community by restoring hope, promoting justice, building peace and providing leadership through their shared space, which is to be a transformative presence in North Belfast.
>
> *(174 Trust 2018)*

As a safe, "neutral," and shared place, the Trust's programming, employees, and participants work to reterritorialize the fractured geography of sectarianism in north Belfast and beyond. The Trust's inclusive and all-encompassing message of "respect and dignity for all" traverses the gulf between sectarian-divided ethnonational communities. It also embraces members of other faiths, sexual orientation(s), gender identification(s), and citizenship status. When community members who utilize the Trust were asked about the significance of the Trust for the local community, one participant explained:

> It fundamentally changed north Belfast. It helped us make friends across the peacelines. Rare friendships like that take time. The Trust gives us the time, the space, and the opportunities to open up and share ourselves. Through those experiences we gain skills to help work for cooperation and reconciliation in our communities.
>
> *(personal interview at 174 Trust, 2017)*

Indeed, research suggests that "cross-group friendships" reduce prejudice and can foster additional trans-ethnonational friendships (e.g., Dowler and Ranjbar 2018).

In recognition of the Trust's achievements, Bill Shaw has been invited to speak about his organization throughout Europe, Canada, and the United States. During a personal interview, Shaw explained that his motivation for establishing the Trust began as a spiritual journey that centered on individual lives:

> It was my conviction that being part of the church was to do God's work, but in Northern Ireland, I realized that meant the church must also have a message of reconciliation. That is the solution. The fact that one of the Trust's Youth Footballer's dad is an active member of the UDA and yet his boy remains friends with his Catholic friends from the football team is evidence, to me, of our success. The friendships that started here now continue outside of the Trust. It's the "starfish story" – you know – when the old man tells the child who is throwing starfish on the shoreline back in the ocean so they won't die – "There are too many to save, throwing them back won't

> make a difference." But the child replies, "It makes a difference to this one." That's the Trust. We make a difference one life at a time. We provide hope for generational change. I know the work is slow, at times glacial, but I also know it is the progress that changes lives. And that, in turn, will begin to change other peoples' lives as well.
>
> *(Shaw, personal interview at 174 Trust, 2015)*

Exclusive ethnonational territoriality is part of everyday life for many in north Belfast. However, while space can be the origin of conflict, the Trust sheds light on how shared space and specifically designed programming can foster cross-communal collaboration, friendships, and steps toward reconciliation. Indeed, it demonstrates how local grassroot efforts can combat the territorial consequences of sectarian divisions in north Belfast and beyond.

This chapter presented a local case study of how people utilize space to foster peace. The following chapter returns to the investigation of Dublin's Garden of Remembrance (Chapter 1) to examine how official national space is employed in reconciliation efforts, renegotiation of national identities in Ireland, and British–Irish geopolitical relations.

References

174 Trust. 2018. "174 Trust Newsletter and Mission Statement." *174 Trust*. www.174trust.org/home-horizon.

Bleakney, Judith, and Paul Darby. 2018. "The Pride of East Belfast: Glentoran Football Club and the (Re) Production of Ulster Unionist Identities in Northern Ireland." *International Review for the Sociology of Sport* 53 (8): 975–96.

@BorderIrish. 2019. *I Am the Border, So I Am*. New York, NY: HarperCollins.

Bryman, Alan. 2016. *Social Research Methods*. Oxford: Oxford University Press.

Campbell, John. 2017. "Belfast Home to Half of NI's 100 Most Deprived Areas." *BBC News*, 2017. www.bbc.com/news/uk-northern-ireland-42103506.

Corrymeela. 2018. "Corrymeela Community." www.corrymeela.org/.

Cunningham, Niall, and Ian Gregory. 2014. "Hard to Miss, Easy to Blame? Peacelines, Interfaces and Political Deaths in Belfast during the Troubles." *Political Geography* 40 (May): 64–78. https://doi.org/10.1016/j.polgeo.2014.02.004.

Dempsey, Kara E. 2022. "Fostering Grassroots Civic Nationalism in an Ethno-Nationally Divided Community in Northern Ireland." *Geopolitics*: 1–17. https://doi.org/10.1080/14650045.2020.1727449.

Dowler, Lorraine, and A. Marie Ranjbar. 2018. "Praxis in the City: Care and (Re) Injury in Belfast and Orumiyeh." *Annals of the American Association of Geographers* 108 (2): 434–44.

Enright, Robert. 2013. "10 Years of Forgiveness Education in Northern Ireland." *International Forgiveness Institute*. https://internationalforgiveness.wordpress.com/2013/01/03/guest-blog-10-years-of-forgiveness-education-in-northern-ireland-2/.

Flint, Colin. 2004. *Spaces of Hate: Geographies of Discrimination and Intolerance in the USA*. East Sussex: Psychology Press.

Gaffikin, Frank, Chris Karelse, Mike Morrissey, Clare Mulholland, and Ken Sterrett. 2016. *Making Space for Each Other: Civic Place-Making in a Divided Society*. Belfast: Queen's University Belfast.

Graham, Brian, and Catherine Nash. 2006. "A Shared Future: Territoriality, Pluralism and Public Policy in Northern Ireland." *Political Geography* 25 (3): 253–78. https://doi.org/10.1016/j.polgeo.2005.12.006.

Harbison, Jeremy. 2002. *Review of Community Relations Policy*. Belfast: Community Relations Unit (CRU).

Hirsch, Marianne. 2008. "The Generation of Postmemory." *Poetics Today* 29 (1): 103–28.

IFI. 2018. "International Fund for Ireland 2018." https://internationalfundforireland.com.

Kasirova, Diloro. 2014. "Implementation of Post-Conflict Reconstruction and Development Aid Initiatives: Evidence from Afghanistan." *Journal of International Development* 26 (6): 887–914.

Lloyd, Katrina, and Gillian Robinson. 2011. "Intimate Mixing – Bridging the Gap? Catholic-Protestant Relationships in Northern Ireland." *Ethnic and Racial Studies* 34 (12): 2134–52.

McAuley, James W., and Neil Ferguson. 2016. "'Us' and 'Them': Ulster Loyalist Perspectives on the IRA and Irish Republicanism." *Terrorism and Political Violence* 28 (3): 561–75.

McDowell, Sara, Máire Braniff, and Joanne Murphy. 2017. "Zero-Sum Politics in Contested Spaces: The Unintended Consequences of Legislative Peacebuilding in Northern Ireland." *Political Geography* 61: 193–202.

Mesev, Victor, Peter Shirlow, and Joni Downs. 2009. "The Geography of Conflict and Death in Belfast, Northern Ireland." *Annals of the Association of American Geographers* 99 (5): 893–903. https://doi.org/10.1080/00045600903260556.

NILT. 2021. "Northern Ireland LIFE & TIMES." *Northern Ireland Life and Times*. www.ark.ac.uk/nilt/.

Nolan, Paul. 2017. "Two Tribes: A Divided Northern Ireland." *The Irish Times*, April 1, 2017. www.irishtimes.com/news/ireland/irish-news/two-tribes-a-divided-northern-ireland-1.3030921.

Williams, Isobel. 2018. *Shoulder to Shoulder*. Film. BT Sports Films.

8

HOPE FOR THE FUTURE II

Employment of the Garden of Remembrance in the Republic of Ireland

This book began with a brief description of Queen Elizabeth and Irish President Mary McAleese's wreath-laying ceremony in the Republic of Ireland's Garden of Remembrance in 2011. This chapter returns to this geopolitically significant space and ceremonial observance to examine how commemorative spaces are employed to reimagine intrastate and British–Irish relations (e.g., McDowell and Braniff 2014). It also provides insight into changing constructions of statehood, national identity, and memory in Ireland.

The highly anticipated Garden of Remembrance, dedicated to "those who died fighting for the Irish nation" and strategically located in the heart of the Irish capital, opened to the public amidst great celebration on Easter Monday, April 10, 1966 (Hanly 1966). Its opening coincided with the fiftieth anniversary of the 1916 Easter Rising. In addition to the general public, officials, and army officers, many who attended the opening ceremony participated in the Rising. The desire for a national commemorative place is not unique. The creation of national monuments is often a manifestation of a deeply felt need to foster national identity and perpetuate the memory of particular events, ideals, individuals, or groups that are considered worthy of remembrance (e.g., Tsang and Woods 2014). This selective process of memorialization and representations of place has been historically transcribed on the landscape through rituals in symbolically significant spaces in Ireland (e.g., Grayson and McGarry 2016). Commemorative urban spaces can hold great meaning in that they may possess certain spatial and political implications that can influence societal understanding of belonging, particularly during times of heightened reflectivity, which can occur during festivals, celebrations, and spectacles.

While the urban situation of these spaces can contribute to memory, urban space can also be seen as "a mnemonic device itself, even as an actor in the process

DOI: 10.4324/9781003141167-8

of shaping memory" (Bădescu 2019, 184). As the Irish Minister of Culture, Jimmy Deenihan TD explained in 2015 during a personal interview:

> The Garden of Remembrance holds great emotional meaning and historical connection for Irish people, built upon national sacred ground. The space honors the sacrifice of all of those who fought for Ireland's freedom, and its presence ensures that their memory and actions will always be remembered.
> *(personal interview, Dublin 2015)*

Located in the heart of the Republic's capital city on a site that was central to the 1916 Rising, the Garden is a symbolic urban space designed to visibly legitimize and celebrate a young Irish republic that was not established until the end of the 1940s, despite a long struggle against British rule. Indeed, the Garden of Remembrance, constructed in a highly symbolic political and urban location, was intended to be a particularly evocative space to express ideals of national permanence and legitimacy, while fostering a sense of shared history and national identity. It is a place where efforts to shape official discourses about the Republic and selected visions of its past are publicly conveyed. It was designed to project an image of the Republic signifying a strong and fruitful future that accordingly honored its long struggle for independence. However, the public has not always passively accepted how the Republic is represented or employed in this national sacred space. In some cases, they have been directly challenged.

This chapter examines the shifting struggle to shape Irish national identity through highly selected official understandings of the past. More specifically, it investigates some of the ways in which the Garden of Remembrance's (hereafter the Garden) symbolic design and strategic employment was key to the promotion of an evolving sense of collective Irish national identity and geopolitical relations with the UK. This chapter begins with a comparison – a brief description of the Irish National War Memorial, a garden that honors Irish WWI soldiers – in order to examine the nation's hesitancy to commemorate these individuals and this war. It compares the Republic's reluctance to the nation's emphasis on the Garden by providing relevant historical context and motivations for its creation. This chapter then explores the strategic employment of this national sacred space and the Garden's symbolic design by focusing on the deliberate inclusion of certain elements of Irish culture, language, religion, and history. This chapter also explores controversies regarding the use and access to this space. Finally, it concludes with a discussion of how this urban space is utilized as a platform for the renegotiation of Irish nationalism and the Republic's geopolitical relationship with the UK.

State efforts to shape national memory

National memory is venerated in various ways, including transcription of the symbolic landscape through monuments and memorials. As national memory is often underpinned by a perceived shared past that emanates from calculated state efforts, "the state is usually the official arbitrator of public commemoration and, therefore, of national identity, and . . . planning, maintaining and funding memorial monuments, programs and events" (McDowell 2008, 40–41). Since national memory and identity are dynamic, the symbolic meaning of corresponding memorials is manipulated, negotiated, and challenged over time.

The creation and function of war memorials is commonly an intrinsic element of nation-building efforts to forge an "imagined community" and "collective forgetting or amnesia" (Johnson 1995, 55). Thus, examination of the "silences" in official national memory can be illuminative as elements that are emphasized. For example, as discussed in Chapter 3, in the 1920s, the newly formed Irish Free State grappled with its past as it worked to forge a national identity. This process advanced through efforts to shape new national narratives, including constructing a national identity that was distinct from rival "British" identity (e.g., Todd 2018; McCarthy 2012). Subsequently, the Irish Free State was reluctant to recognize its historical cooperation with Britain, including its participation in WWI under the British flag.

Indeed, Ireland contributed to WWI as part of Britain, fighting in all substantial battles including the Somme, Messines, and Gallipoli. The island of Ireland lost 35,000–50,000 individuals in the war. However, as discussed in Chapter 2, during this war, Irish republicans declared independence from Britain in the 1916 Easter Rising and by 1922, the state officially partitioned the island. These political divisions fostered contrasting spheres of remembrance. While unionists and loyalists venerated WWI war sacrifices, many Irish republicans were reluctant to commemorate this war. In many ways, recognizing Irish involvement in WWI conflicted with republicans' emerging national narrative, particularly for those who wanted to exclude any association with the state and its imperialist past.

During the IWI/AIW, the popular Irish rebellion song, Foggy Dew, honors the Rising participants and insinuates that Irish who answered Redmond's call to fight in WWI should have stayed in Ireland to support the Rising instead. As two of its verses suggest:

> Right proudly high over Dublin town they hung out the flag of war
> 'twas better to die 'neath an Irish sky than at Suvla or Sud-El-Bar
> And from the plains of Royal Meath strong men came hurrying through
> While Britannia's Huns, with their long-range guns sailed in through the foggy dew
> 'Twas England bade our wild geese go, that "small nations might be free"
> Their lonely graves are by Suvla's waves or the fringe of the great North Sea.
> Oh, had they died by Pearse's side or fought with Cathal Brugha
> Their graves we'd keep where the Fenians sleep, 'neath the shroud of the Foggy Dew.
>
> *(Foggy Dew/Down in the Glen lyrics attributed to Canon Charles O'Neill, Greaves 1980)*

The lyrics imply it is better for Irish to die in Ireland fighting against the British than dying at Suvla (Sulva Bay) or the British occupied Dardanelles fort, or at Sud-el-Bar, in the ruinous Gallipoli campaign in Turkey (1915–1916). The song admonishes these WWI soldiers, suggesting that instead of fighting alongside republican revolutionaries such as Patrick Pearse or Cathal Brugha – which would have earned them a place of honor in national graves in Ireland "under the Foggy Dew" – those who died in WWI are lying in "lonely graves" in Turkey.

Commemorating Irish who fought in WWI: the National War Memorial, Islandbridge

What began as a nebulous hesitancy to officially recognize Irish contributions in WWI eventually became a notable silence in Irish history. While the Irish Free State's administration permitted memorial services and parades on Armistice Day to celebrate the end of WWI, it did not endorse or support them. This section analyzes the National War Memorial as an example of this official reluctance.

When members of the public asked the government of the Irish Free State to erect a war memorial to commemorate WWI soldiers in the 1920s, subsequent discussions regarding its necessity, symbolism, and geographic location became contentious. While the president of the Free State, W.T. Cosgrave, supported the project, many republicans challenged the appropriateness of the proposed memorial space. Some argued it would be erroneously interpreted as a monument to the British Empire. Others rejected its initial proposed location in Dublin's centrally located Merrion Square, due to its close proximity to the seat of government, which detractors argued could signal the government's sanction for the project. Indeed, the Irish Minister of Justice, Kevin O'Higgins (1922–1927), insisted its location should be out of public sight, and therefore, out of mind (Byrne 2014).

Eventually, the government approved a plan in 1931 to construct a monument outside Dublin's city center, in the Islandbridge suburb. The decision to place the National War Memorial outside Dublin is significant. Its appointed location stands in stark contrast to the Garden of Remembrance's desirable site in the symbolic heart of the capital city (see Map 8.1).

When construction of the National War Memorial concluded in 1939, Taoiseach Éamon de Valera postponed its official opening due to the imminent outbreak of WWII. After the Republic of Ireland Act 1948 severed the official connection between the British Commonwealth and the Free State in 1949, the leadership of the Irish Republic did not prioritize commemorating the National War Memorial. During this same time, the national school board removed detailed discussions of WWI from national schoolbooks and student curriculums (e.g., D'Arcy 2007; Byrne 2014).

In 1956 and 1958, republican paramilitaries bombed the National War Memorial, and without state financial investment, it eventually became a city dumping site utilized by Dublin City Refuge Disposal (Myers 2014). As part of an urban renewal project in the mid-1980s, the state restored the War Memorial and opened it to the public in 1988 (see Figure 8.1). However, no representative of the Irish

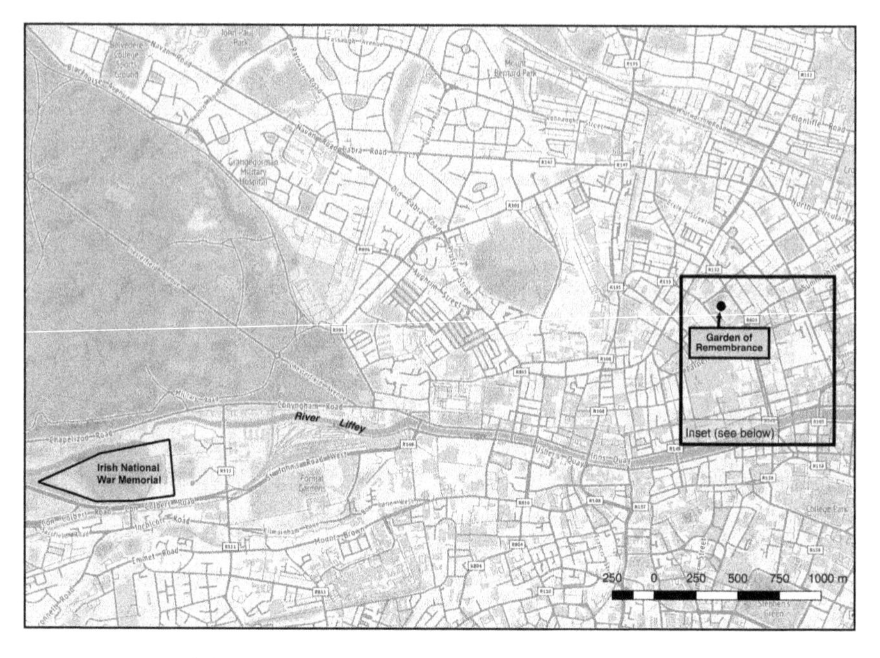

MAP 8.1 Map of the Garden of Remembrance Memorial and Irish National War Memorial

FIGURE 8.1 Irish National War Memorial, Islandbridge, Republic of Ireland.

Source: Photo by author.

government attended the official commemoration. In essence, the state's reluctance to recognize Irish involvement in WWI fostered a strategic "disremembering" of the event and silenced it in officially promoted Irish national memory. Indeed, what is nationally commemorated is highly selective:

> as competition over commemoration of trauma in Ireland has been evident in rivalries between those who choose to wear a poppy around the month of November, initially commemorating the dead of the Great War and by extension all those who have since died in the service of the United Kingdom and the British Empire, and those who wear a lily at Easter time, commemorating the dead of the Rising and by extension all those who have since died for the republican cause.
>
> *(Morris 2005, 140)*

Designing national memory: the Garden of Remembrance, Dublin

In contrast to Irish involvement in WWI and the National War Memorial, there were many who wanted to venerate the nascent Irish nation visibly on the landscape. As early as 1935, the Irish Army's Dublin Brigade Council called for the creation of a memorial to honor those who fought for the Irish nation. The first concrete plans, designed in 1961, placed a memorial in the center of Dublin.

The Garden of Remembrance's central location reflects the importance of the monument, not only to the Irish struggle for independence but also to the history of the city as a significant site during the Rising. Situated just north of the General Post Office (GPO), one of the most important garrison locations during the 1916 Rising and the location from which Pearse first read the Proclamation of the Irish Republic, the monument honors one of the pivotal places in the unfolding drama of the rebellion. The Garden is also located within the greater Rotunda Gardens, the site of the founding of the Republican Irish Volunteers, many of whom participated in the 1916 Rising. Additionally, this commemorative space is constructed on the location where several of the leaders of the Rising were kept overnight before they were officially jailed. Many of these leaders were subsequently executed, triggering a chain of events that led to great turmoil, a growing sense of Irish nationalism, and republican sympathy among much of the Irish public. Participants and "martyrs" of the Rising quickly became national heroes and revered in classroom textbooks authorized by the Department of Education (McCarthy 2012). Since the Garden of Remembrance is dedicated to those who died fighting for an independent Ireland, it is also not surprising that this national commemorative space is strategically located just off O'Connell Street (formerly Sackville Street), named in honor of Ireland's "Liberator," and along Parnell Square, named for Charles Stewart Parnell, who fought for Irish legislative independence (see Map 8.2).

MAP 8.2 Inset: Location of the Garden of Remembrance in Dublin

While the location is significant in relation to Irish national events in the first decades of the twentieth century, the Garden also commemorates other key events, including Wolfe Tone's 1798 Rebellion and the IWI/AIW. Accordingly, the location and the scope of the Garden transform an ordinary urban space into a commemorative memorial place that honors the nation. In this way, the Garden plays an active role in the tribute and creation of national narratives about the Republic of Ireland's struggle for national independence, identity, and national legitimacy.

Dedication and design

On Easter Monday 1966, President Éamon de Valera, a participant in the rising and a signatory of the Proclamation of the Irish Republic, presided over the Garden of Remembrance's opening ceremony. The Garden's public début, which marked the fiftieth anniversary of the Rising and the Proclamation of Irish Republic, was part of a larger celebration that included a military parade and a commemorative celebration at the GPO. Many attended its opening, including hundreds who participated in 1916 events, Irish army officers, government officials, and members of the general public. One of the attendees, who had served in the GPO garrison during the Rising, was carried into the Garden for the ceremony on a stretcher and greeted with applause by onlookers.

Highlighting the intended purpose of the Garden, a local reporter called it a "national shrine" and "a place of pilgrimage that attracted Irish young and old alike" (Irish Press 1966a). The intentional use of the word "pilgrimage" not only suggested that the Garden was to become a site of such great national significance that people would travel to see it but also refers to the importance of religion in conceptions of Irish nationalism and national identity. The Catholic Church long played a role in Irish politics and culture – the Constitution of the Republic of Ireland included the declaration, removed only in 1973, that the "State recognizes the special position of the Holy Catholic Apostolic and Roman Church as the guardian of the Faith professed by the great majority of the citizens" (Constitution of Ireland 1937).

In order to emphasize the Catholic Church's "special position" within the Republic, the Archbishop of Dublin, Reverend John Charles McQuaid, was invited to bless the memorial gardens at the beginning of the opening ceremony. At the conclusion of the Archbishop's blessing, President de Valera addressed the audience by speaking first in Irish, the national and first official language of the Republic of Ireland, about the importance of commemorating those who fought and died for the Irish nation (Irish Press 1966a). In essence, the opening ceremony worked to consecrate the Garden of Remembrance as a sacred, national space for Ireland.

The prominent role of the Catholic Church, the strategic use of the Irish language, and the celebration of past historical events served to create a sense of national unity and collective national identity. However, they also highlighted and heightened key symbolic design elements that evoked aspects of Irish religion, Irish language, culture, and history. Architect Daithi Hanly's design for the Garden had been selected for its highly historical-symbolic elements that commemorated Irish sacrifice and transcendence. Hanly believed the site was a symbolic place of execution and therefore wanted to create a place that both demonstrated and celebrated a narrative of sacrifice, suffering, and redemption (see Figure 8.2). His intentional invocation of national martyrs reflects his desire to create a design that would remind visitors of the numerous sacrifices made by the Irish in their centuries-long struggle for an independent Ireland (Irish Press 1966b).

Other elements of Hanly's design for the Garden of Remembrance capture this theme. Its overall shape is rectilinear with a sunken cruciform shape, reminiscent of a Latin-cross-style church, carved out of its center. Within this sunken cruciform shape, there is a raised central platform with the national flag. The platform serves as a national altar, designed with a pedestal for wreaths or an important speaker, while the centerpiece of the Garden is a reflecting pool, set in the center of the sunken area – also in the shape of a Latin cross. The remaining raised lawn is seven feet higher than this pool, in dramatic contrast to the sunken cruciform shape where visitors walk to the Garden. The geometric form of the memorial garden suggests a rigid military character and discipline within the space. Additionally, the cruciform shape and the sunken part of the Garden combine to set a solemn and religious tone within the otherwise busy city block. In this way, the Garden

FIGURE 8.2 Sunken gardens, Garden of Remembrance, Dublin, Republic of Ireland.

Source: Photo by author.

connects the sacred with the secular in material form to create a special place that honors and legitimizes the Irish nation.

In order to enter the Garden, which is enclosed by external walls, visitors must pass through one of two dramatic entrances, thus creating a clearly defined urban space. The entrances feature 50-foot-wide sliding gates and large letters that declare in both Irish and English that the Garden is "dedicated to those who gave their lives in the cause of Irish freedom" (Hanly 1966). The use of Irish is also evident throughout the Garden, particularly on one of the walls that displays author Liam Mac Uistín's poem, *We saw a vision*. This poem, which was added two decades after the Garden first opened to the public, focuses on Ireland's desire for freedom, asking visitors to be mindful of those who died and the sacrifices they made fighting for the establishment of the Republic. The word *saorise* (freedom) was employed often during the Irish struggle for independence and remains a dominant term in republican neighborhood murals in Northern Ireland today. With regard to the importance of the Irish language to national culture and identity, it is significant to note that Uistín's poem was originally written only in Irish. Translations in English and French were added to the site only in 1976 so that more visitors would be able to read the poem.

While the reflecting pool's shape is highly religious in nature, the symbols included in its design highlight some of the historical and cultural elements of Ireland's past. For example, the bottom of the pool is decorated with green and blue mosaic tiles in a wave pattern punctuated with a design featuring daggers, broken spears, and bright yellow shields. The appearance of broken weapons at the bottom of the pool references the Celtic tradition in which war parties broke their weapons and tossed them into rivers at the end of a conflict to symbolize the conclusion of hostilities (see Figure 8.3).

FIGURE 8.3 Mosaic of Celtic post-war traditions, Garden of Remembrance, Dublin.

Source: Photo by author.

Additionally, the railings that flank the sunken cruciform area include important Irish emblems modeled from cultural artifacts held at the National Museum, such as the national Irish harp and a 2,000-year-old Irish sword. However, they also include religious emblems, such as the Cross of Cloyne – an early Irish Christian Cross – in addition to wrought iron bars that represent eternal light, again demonstrating the close relationship that existed between Irish culture, history, and religion (Hanly 1966).

The focal point of the Garden features the Irish national flag in the center of the raised platform with a pedestal for wreaths. The Irish flag is centered between all four provincial flags, representative of the four historic Irish provinces, which flank the main entrance on the opposite side of the Garden. The inclusion of these four flags is symbolic of a still-unsatisfied aspiration for a united Ireland. As discussed in Chapter 1, while all counties in three of the four historic provinces of Ireland (Connacht, Leinster, and Munster) are presently located within the Republic of Ireland, only three of the nine counties of the fourth province (Ulster) are in the Republic. Northern Ireland comprises the remaining six counties and remains a source of great strife for many in the Republic and Northern Ireland today.

In 1971, the state added Oisin Kelly's rather contentious sculpted centerpiece to the raised platform in front of the Irish flag. Kelly's highly emotional and provocative sculpture of the Irish myth of the Children of Lir was also inspired by a famous line in William Butler Yeats' poem, "Easter, 1916." Yeats describes those who were involved in the 1916 Rising as "transformed utterly" and subsequently for an Ireland in which, "A terrible beauty is born" (Yeats 1920). The statue relates to Yeats' description through the depiction of Lir's four children in the midst of transforming into swans (see Figure 8.4).

According to the Irish myth, their jealous stepmother turned Lir's children into swans for 900 years, mirroring Ireland's almost 900-year struggle against British control (Irish Press 1966a). While several variations of this legend exist, they commonly conclude with the children's conversion back to human form – as a result of a Christian blessing, the sound of a church bell, or a baptism. By incorporating the religious and transformational elements of this legend, the statue is symbolically representative of the Irish nation and its rebirth as the modern Irish Republic with Irish myths, history, and Christianity. Nonetheless, the statue was not universally embraced. Some government officials expressed concern regarding whether the sculpture was appropriate in this national space. However, after searching for alternative designs, most agreed that "it would be extremely difficult, if not impossible, to find a more suitable alternative" (Irish Press 1966b).

The prominence of Irish myths and historical artifacts in the Garden is significant. Even the key fashioned for the entrance gate was designed as a replica of the oldest known key in Ireland. This intentional effort to emphasize the past was motivated by a specific politically driven, anti-partition political campaign championed by President de Valera in a variety of speeches delivered in international fora between 1948 and 1951 protesting the 1921 Anglo-Irish Treaty. One of his central arguments against what he considered to be an illegal division of the island was that Northern Ireland – a nonhistorical creation – divided the "ancient nation

FIGURE 8.4 Statue of transformation of the Children of Lir, Garden of Remembrance, Dublin.

Source: Photo by author.

of Ireland" (Kelly 2011). Thus, for strategic as well as celebratory reasons, this symbolic memorial space seeks to draw connections to Ireland's past, its language, history, culture, and religion. More than a space to commemorate national martyrs, the Garden works to legitimize the creation of the Republic of Ireland and the purported need to reunite the island under the Republic's flag.

A selective space

Given the polyvocal nature of state-building projects, not all individuals will necessarily agree with the officially promoted meaning or representations. It is important to recognize that the official iconography and meaning of national and commemorative landscapes are subjective entities open to contestation or subversion (e.g., Hagen and Ostergren 2020). Public reaction to Hanly's Children of Lir sculpture, for example, ranged from warm acceptance of the notion that the romanticism portrayed in the statue was evocative of the spirit of those who died for Ireland, to ridicule for depicting what some considered an old myth that was not about resurgence or transcendence but death and loss instead (Hanly 1966).

Access to this national space and how the government determines its use also are significant. From its inception, the Garden was a key location for commemorative marches and rapidly became a stage for political rallies and protests. For example, thousands of protesters marched there to express their shock and horror after Bloody Sunday (1972) in Londonderry/Derry (McDonald 1973). The intentional incorporation of the Garden during this protest demonstrates protesters' efforts to gain legitimacy and gravitas by drawing on the power and significance endowed in this sacred, national memorial space. The Garden has also served as the political platform from which Taoiseach Jack Lynch publicly advocated for reconciliation and a peaceful reunification of the island, a statement that upset many within the British government (Irish Press 1971). Some IWI/AIW veterans have chosen the Garden as the venue for their annual outings and reunions (Irish Times 1972). The Garden has even been the location for a protest staged against the Republic's acceptance into the European Economic Community (EEC) on the grounds that Ireland was entering without the six counties of Northern Ireland.

Moreover, not all have been welcomed into this urban space. As in many cases with memorial spaces that commemorate a historical or political past, utilization and access can become a source of contention within the community. For example, concerts were held regularly in the Garden until banned in 1967 as a result of complaints by the IRA's Council of the Dublin Brigade. The council demanded that the Garden only be used to honor those who died for Ireland, insisting that "the Garden of Remembrance be maintained in the spirit in which it was conceived, as an honour to those who made the supreme sacrifice for Ireland" (Irish Press 1967). Additionally, plans to host the first National Commemoration Day at the Garden, which would have honored all Irish who died in past wars, including Irish who died during WWI, were altered due to protests. Instead, the government hosted smaller ceremonies at various denominational churches throughout Dublin (Coghlan 1987).

Contentious debates pertaining to access to the Garden also occurred around societal proceedings. For example, the Irish National Gay Federation's efforts to host a wreath-laying ceremony in the Garden "in memory of the many millions of persons, who, over the centuries have been tortured, imprisoned or killed because of their sexual orientation" was blocked by the national government, which argued that the "stated objectives of the ceremony were not in keeping with the purpose for which the garden was dedicated" (Delaney 2013). While the government affirmed that it supported the need to commemorate victims of abuse, it made clear that it did not believe the Garden of Remembrance was the appropriate location for this event.

Another debate regarding the access to and the use and meaning of this national space occurred during Queen Elizabeth's visit to the Republic in May 2011. Her visit was to commemorate the 1998 Good Friday/Belfast Agreement, a major advance in the peace effort for Northern Ireland. Irish President Mary McAleese, who was born in Belfast and lived in the city during the Troubles, was a staunch supporter of reconciliation and eager for the queen's visit. While the 4-day visit to the Republic was generally considered a success by the Irish media and the public, Sinn Féin publicly objected to the queen's presence, stating that the time was not right for a British monarch to visit the Republic. It is therefore not surprising that the most controversial event during the queen's trip was her visit to the Garden.

Sinn Féin and members of the IRA vehemently opposed her presence in this national sacred place and promised to organize protests to block her entrance to the Garden. Many unionists and loyalists in Northern Ireland were also uncomfortable with the queen visiting a memorial garden that honors those who fought against the British. Fearing potential disturbances, the Irish government restricted access to the ceremony and closed the streets surrounding the entrance, including that of the local Sinn Féin headquarters on Parnell Square. As a result, the queen's motorcade traveled along deserted streets to the Garden, and the public was unable to see her pass through its gates. The small protests organized by certain republican groups were confined to the southern and northern edges of Parnell Square.

Commemorative space as a medium for negotiating a troubled past

Arguably, the queen's presence in the Garden was the most politically and culturally significant moment of her visit to the Republic as she, accompanied by Irish President Mary McAleese, walked to the raised platform, laid a memorial wreath on the national altar, and observed a moment of silence. The fact that a British monarch could be publicly welcomed by the Republic's political leaders in a space that honors those who died primarily fighting against what the queen embodies is a demonstration of the Republic's more recent openness to altering century-old political traditions. Additionally, Queen Elizabeth's willingness to enter this Irish national place and honor deceased Irish patriots was a momentous indication of a similar acceptance. While it was generally a silent affair, the actions of these political leaders in this highly politically symbolic space worked to redefine the complicated Irish–British relations. By fostering new perceptions for this exclusionary national space in Ireland, the Garden provided these heads of state a milieu for peacebuilding – a place to create more inclusive and plural societies.

The Republic's Garden of Remembrance has become what the Dublin Brigade Council originally desired – a national place that honors those who died for the Irish nation. However, the Garden has additional meanings and functions within

Irish society. This symbolic place is hallowed ground within the capital's urban core and promotes a sense of national identity, history, and legitimacy. It is a place where negotiations of Irish society are publicly manifested or challenged, because it is a medium through which the interrelations and subjective entities that exist between the state and society are mediated to the larger community and across a range of scales (e.g., Williams and McConnell 2011; McDowell, Braniff, and Murphy 2017). Because iconic national places, including their use and control, are not neutral, such places are constantly open to contestation and alternative interpretations and employment by the public.

As the Republic continues to negotiate its relationship with the UK, while simultaneously struggling with certain historical elements of its own past and issues of contemporary society, this commemorative space serves as a mirror. The queen's visit to a garden that was created to help define a nation, primarily by honoring those who fought against the British, signals a significant potential transformation for the Republic and its relationship with the UK. Many in Ireland hope the queen's visit to the Garden of Remembrance is an indication of a renegotiation of the troubled relationship between the states. In the words of President McAleese, there is hope that these actions will further aid in reconciliation efforts to "forge a new future, a future very, very different from the past, on very different terms from the past" (Irish Times 2011).

The Garden is a platform from which geopolitical relations are renegotiated, the state projects selective national images, and the public challenges the spatial practices that comprise ever-changing constructions of a nation and corresponding identities. The Republic also continues to demonstrate the possibilities of small state diplomacy through its role in Brexit proceedings and its contributions as a member of the EU and the UN Security Council. However, Ireland is at another critical juncture as challenges to the Northern Ireland Protocol threaten Irish–British–EU geopolitical relations, reconciliation efforts in Ireland, and the spatial constitution of political units on the island. The book's conclusion focuses on the geographic and political implications of this debate.

References

Bădescu, Gruia. 2019. "Making Sense of Ruins: Architectural Reconstruction and Collective Memory in Belgrade." *Nationalities Papers* 47 (2): 182–97.

Byrne, Elaine. 2014. "The Forgotten Irish Soldiers Who Fought for Britain in the First World War." *The Guardian*, 2014. www.theguardian.com/world/2014/apr/05/irish-soldiers-who-fought-for-britain.

Coghlan, Dan. 1987. "Haughey to Drop Remembrance Ceremony." *The Irish Times*, 1987.

"Constitution of Ireland, Article 44." 1937. https://www.constituteproject.org/constitution/Ireland_2019?lang=en

D'Arcy, Fergus. 2007. *Remembering the War Dead: British Commonwealth and International War Graves in Ireland since 1914*. Dublin: Stationery Office.

Delaney, Eamon. 2013. "National Volunteer's Garden Not the Place for a Child Abuse Memorial." *Irish Independent*, 2013.

Grayson, Richard S., and Fearghal McGarry. 2016. *Remembering 1916: The Easter Rising, the Somme and the Politics of Memory in Ireland*. Cambridge: Cambridge University Press.
Hagen, Joshua, and Robert C. Ostergren. 2020. *Building Nazi Germany: Place, Space, Architecture, and Ideology*. Lanham, MD: Rowman & Littlefield.
Hanly, Daithi. 1966. "Garden Fit for Heroes." *Sunday Independent*, 1966.
Irish Press. 1966a. "Final Selection of 1916 Garden Sculpture: Oisin Kelly Chooses 'The Children of Lir' as Theme." February 28, 1966.
———. 1966b. "Garden of Remembrance." May 19, 1966.
———. 1967. "Protests over Bands in Garden." August 4, 1967.
———. 1971. "Necessity of Reconciliation." 1971.
———. 1972. "IRA Veterans Annual Outing." August 9, 1972.
Irish Times. 2011. "McAleese Hails 'Extraordinary Moment.'" 2011. www.irishtimes.com/news/mcaleese-hails-extraordinary-moment-1.876601.
Johnson, Nuala. 1995. "Cast in Stone: Monuments, Geography, and Nationalism." *Environment and Planning D: Society and Space* 13 (1): 51–65.
Kelly, Steven. 2011. "A Policy of Futility: Eamon de Valera's Anti-Partition Campaign, 1948–1951." *Etudes Irlandaises* 26 (2).
McCarthy, Mark. 2012. *Ireland's 1916 Rising: Explorations of History-Making, Commemoration & Heritage in Modern Times*. London: Ashgate.
McDonald, John. 1973. "Bloody Sunday Commemoration." *Irish Press*, 1973.
McDowell, Sara. 2008. "Heritage, Memory and Identity." In *The Ashgate Research Companion to Heritage and Identity*, 37–54. Farnham: Ashgate Publishing.
McDowell, Sara, and Maire Braniff. 2014. *Commemoration as Conflict: Space, Memory, and Identity in Peace Processes*. Basingstoke: Palgrave Macmillan.
McDowell, Sara, Máire Braniff, and Joanne Murphy. 2017. "Zero-Sum Politics in Contested Spaces: The Unintended Consequences of Legislative Peacebuilding in Northern Ireland." *Political Geography* 61: 193–202.
Morris, Ewan. 2005. *Our Own Devices: National Symbols and Political Conflict in Twentieth-Century Ireland*. Dublin: Irish Academic Press.
Myers, Kevin. 2014. *Ireland's Great War*. Dublin: Lilliput Press.
Todd, Jennifer. 2018. *Identity Change after Conflict: Ethnicity, Boundaries, and Belonging in the Two Irelands*. London: Palgrave Macmillan.
Tsang, Rachel, and Eric Woods. 2014. *The Cultural Politics of Nationalism and Nation-Building*. New York, NY: Routledge.
Williams, Patricia, and Fiona McConnell. 2011. "Critical Geographies of Peace." *Antipode* 43 (4): 927–31.
Yeats, William Butler. 1920. *Easter, 1916*. Vol. 25, 69. New York, NY: The Dial.

Note: The majority of this chapter appears in Dempsey, Kara. 2018. "Creating a Place for the Nation in Dublin." *The City as Power: Urban Space, Place, and National Identity* 27-40. Washington: Rowman & Littlefield. Used with permission.

CONCLUSION

Renegotiating national memory and Brexit: an uncertain future in Ireland

This book attempted to elucidate the profound and cumulative means in which space is a fundamental and transformative element of the production and (re)negotiation of ethnonational territorial conflicts, bordering processes, identity construction, and peacebuilding in Ireland. As Irish geographer Nuala Johnson has argued,

> Space is more than the container in which historical narratives of memory are placed . . . landscape becomes the process of memory construction that involves inscription and erasure, consensus, and conflict, reflection, and action in the public performance of remembrance.
>
> *(2003, 171)*

Indeed, space is an integral part of statecraft, nation-building, and – as this book explores – the repositioning of geopolitics in Ireland that is occurring both at different political scales and across various key debates.

The Republic of Ireland

The Republic of Ireland has become an active force in a globalizing world, demonstrating an expanding role for small state diplomacy on the international stage. Examples of Ireland's newfound activism are many. For example, former Irish president Mary Robinson served as UN High Commissioner for Human Rights from 1997 to 2002. The state is an active member of the EU and its political leadership has served as the President of the Council of the European Union six times. In 2020, Irish politician Mairead McGuinness was appointed as the EU's European Commissioner for Financial Services, Financial Stability, and Capital Markets Union. In this key role, the Republic's representative is tasked with "preserving

DOI: 10.4324/9781003141167-9

financial stability, protecting savers and investors, fighting financial crime, as well as ensuring the flow and access to capital for businesses and consumers in the European Union" (Europa.eu/European-union 2021). McGuiness' goal is to develop the EU's single market as a globally competitive force and strengthen the value of its common currency, the Euro.

In 2021, the Republic joined the UN Security Council. Through this new role, the state can further its contribution to peace and security on a global scale. As the Irish Minister for Foreign Affairs explained:

> The United Nations is at the heart of Irish foreign policy. Whether it is through our unbroken record of UN peacekeeping since 1958 or our leading roles in disarmament and non-proliferation, we know what can be achieved when countries work together. We take our seat on the Council at a very challenging time, but we are determined to play our part to build the trust and political will necessary to achieve progress in even the most intractable conflicts . . . Membership of the UN Security Council is an opportunity for Ireland to make a significant international contribution, to strengthen our relations with key partners, and to project our values on the global stage. It is in keeping with our long-standing tradition of an independent and principled foreign policy and support for the UN.
>
> *(Dfa Press Release 2021)*

The Republic of Ireland's emerging active international role is a reflection of a society that is increasingly diverse, progressive, and outward-facing (e.g., McGinnity et al. 2018). As Michael Higgins, speaking of the Republic's future in his 2011 presidential acceptance speech stated:

> James Connolly once said, "Ireland without her people means nothing to me." Connolly took pride in the past but, of course, felt that those who excessively worshipped that past were sometimes seeking to escape from the struggle and challenge of the present. He believed that Ireland was a work in progress, a country still to be fully imagined and invented – and that the future was exhilarating precisely in the sense that it was not fully knowable, measurable. The demands and the rewards of building a real and inclusive Republic in its fullest sense remains as a challenge for us all, but it is one we should embrace together.
>
> *(rte 11.1.11)*

As the Republic of Ireland continues to negotiate its geopolitical relationships with the EU and the world at large, it simultaneously struggles with the UK, with certain historical elements of its past, and its diversifying contemporary society. The Republic's commemorative Garden of Remembrance, discussed at the beginning and end of this book, serves as a mirror through which one can view the construction and contestation of elements that constitute the modern Irish nation. As we have seen, the Garden was created to legitimize a nascent nation by honoring those who fought Britain for an independent republic. However, the Queen's 2011 visit to the Garden, at the invitation of Irish President McAleese, signaled a significant geopolitical and social repositioning for the Republic, its national identity, and its relationship with the UK. Indeed, during the Queen's visit to the Republic, she expressed remorse regarding their tumultuous shared past. In a speech delivered at a state banquet in Dublin, the Queen declared:

> It is a sad and regrettable reality that through history, our islands have experienced more than their fair share of heartache, turbulence, and loss . . . with the benefit of historical hindsight, we can all see things which we wish had been done differently, or not at all.
>
> *(cited in Anderson 2016)*

Many hope such statements are an indication of a renegotiation of the troubled relationship between these states, one that no longer highlights competing territorial claims. Instead, these efforts foster new ways to remember the past. Indeed, this ceremony was part of a government initiative to highlight a shared past; one that could help shape a new, more inclusive national memory.

Re-remembering a shared past – WWI commemoration in the Republic

Beginning in the 1990s, the Irish government began to reincorporate its participation in WWI into Irish national memory and forge a new political practice of shared remembrance. In 1993, Irish President Mary Robinson became the first Irish head of state to attend a WWI remembrance service at an Anglican church. That same year, the state launched the *Decade of Centenaries (1912–1922)* commemoration project. In 2012, the National Library of Dublin opened a WWI exhibit that explored Irish contributions during the war. In 2014, the state organized a series of well-attended events (including by many British political leaders) that honored the sacrifices of WWI. Significantly, the government hosted these events in symbolic and sacred places in Dublin, including Glasnevin Cemetery, the resting place for many national heroes (Byrne 2014; Bowman 2014).

The decade-long project aimed to "commemorate each step that Ireland took between 1912 and 1922 in a tolerant, inclusive and respectful way" (www.decadeofcentenaries.com). This more collaborative representation of Ireland's geopolitical relationship with Britain during WWI works to redefine how the Republic remembers and commemorates the war and, "at last, make it okay and not shameful to talk about family members who served in the Allied forces during the Great War" (Bunbury 2020, 201). More specifically, these politically choreographed shared commemorations foster official reinvestigations of Irish national memory and signify a rediscovery of commonalities for divided sectarian communities across Ireland. As Irish and British political leaders highlight elements of a united past in geopolitically significant space, their efforts signify a desire to advance the peace process and forge more inclusive nationalist narratives. Additionally, by supporting local grassroots cross-community centers and several public-engagement programs[1] that contribute to new forms of ritualized reconciliation, the Republic and the UK may be cautiously forging a new future – one that does not underscore territorialized identities conceptualized as dualistic and distinct.

Northern Ireland

Northern Ireland's post-conflict society continues to evolve through volatile processes of peacebuilding that have yet to forge a successful, region-wide cross-community nation-building project. Instead, the internalized dominant paradigm within Northern Ireland continues to assume the presence of two distinct and opposed communities. Even the 1998 peace agreement identifies Northern Ireland's "two communities," a declaration that reinforces perceived distinctions between communities and excludes dynamism and diversity outside the binary categories. Indeed, while the agreement calls for a "fresh start" for the "achievement of reconciliation, tolerance, and mutual trust, and to the protection and vindication of the human rights of all," it does not offer initiatives for creating shared space or a shared Northern Irish national identity.

Many peace organizations argue that protracted sectarianism can only be transcended if people are able to highlight elements of a common past or forge a common future. Indeed, as Chapter 4 demonstrated, residential proximity does not guarantee unity, especially for communities that continue to be divided by peacelines and segregated mental maps. Despite large economic investment in support of cross-community projects in Northern Ireland (e.g., Peace Funds from the EU and International Ireland Fund), evidence suggests that these efforts have, for the most part, only minimally contributed to successful community integration (e.g., Nolan 2017).

While "top-down" cross-community projects in Northern Ireland have experienced only limited success, grassroots engagement and production of shared space have proven effective at the local scale. These efforts can serve as templates for best practices that can be adapted to unique environments to bring various communities together in Northern Ireland's society in transition. However invaluable as these actions may be, the transformative progress is fragile in a society undergoing recent notable demographic changes. For example, earlier census reports for the region that categorizes residents in ethnonational categories suggest that the 2021 census may reveal that, for the first time, Protestants no longer constitute the majority of inhabitants in Northern Ireland[2] (Coakley 2021). In another indication of a shifting demographic balance, Sinn Féin recently surpassed the DUP as the largest party in the Belfast assembly.

1 For example, several Irish universities launched the "Letters of 1916," offering digital archives of Irish letters written from the front during WWI. The public can access these files, search, and add to them.
2 98% of Northern Ireland's population self-identifies ethnically as white (nisra.gov.uk).

Unfortunately, the turmoil generated over the border discussions relating to Brexit and the subsequent Northern Ireland Protocol threatens peacebuilding efforts. Indeed, it was EU funding that financially underpinned many of the organized initiatives to foster community cohesion and reconciliation in Northern Ireland. The EU also facilitated the 1998 peace agreement's demilitarization of the border in Ireland. Thus, the UK's departure from the EU is concerning for many, as peace processes are not static – subject to both progression and the potential for regression. However, while Brexit undermines the EU's supranational political and economic union, it also represents the UK's efforts to reassert the centrality of the state. Clearly, the Republic of Ireland and Northern Ireland are at a critical geopolitical juncture that is currently unfolding across space.

Brexit and the Northern Ireland Protocol – threatening a fragile peace

As discussed in Chapter 1, after the 1998 peace agreement, border checks in Ireland were abolished and the border was demilitarized. However, the future of the Irish border became a prominent point of contention when the UK considered a "full hard exit" from the EU. As the Republic is a member of the EU, the Irish border would become the new, international border for the EU and, subsequently, its customs checks. This insinuates the need to institute a "hard," militarized EU border between Northern Ireland and the Republic to ensure the safety of the supranational union. This suggests separate passport lines at the border that would divide EU (i.e., Irish) and non-EU (i.e., British) citizens. Highlighting these "Irish" and "British" divisions at the border will reignite tensions and violence (e.g., infuriate hardline republicans) that have been relatively dormant since 1998. Fear and trepidation regarding the impending changes have already resulted in a number of casualties.

Recognizing the potentially damaging ramifications, the UK, the EU, and the Republic of Ireland agreed that protecting the 1998 peace agreement was imperative and formulated the expedient of the Irish border "backstop" to maintain the current status of a seamless Irish border. In the case of Ireland, the willingness to modify EU supranational border regulations is indicative of international recognition of the island's tumultuous past. It also demonstrates the potential power that rests within the small island to destabilize larger international and supranational geopolitical relations. Indeed, its role within global geopolitics extends beyond Brexit negotiations to include international asylum and migration discussions, reconciliation efforts in other areas of protracted conflict, and equity rights for minority communities, among many others.

However, loyalist DUP politicians adamantly opposed the proposal because the Irish backstop relocated the EU checkpoints from the Irish border to the coast of Ireland (i.e., the boundary between Northern Ireland and the remaining part of the UK). The DUP, which intensely resists any measures that potentially weaken Northern Ireland's union with the rest of the UK, argued that additional

"Northern Ireland checks only" would alienate and potentially threaten their ties within the UK.

The DUP's refusal to support the Irish backstop was significant as the party's electoral support was critical to former British PM Theresa May's minority party. Without it, May's Brexit proposals were not successful, and she resigned in 2019. Boris Johnson, who succeeded May as British PM in July 2019, called the backstop "a 'monstrosity' that wipes out the UK's sovereignty" and demanded the backstop be removed from the withdrawal deal (cited in McCormack 2019). However, Irish Taoiseach Leo Varadkar insisted that Johnson's plan to eliminate the Irish backstop "would not happen" (Kelly 2019). Ultimately, PM Johnson and the EU signed the Northern Ireland Protocol that came into force on January 1, 2021.

The Northern Ireland Protocol helps prevent border checks from occurring along the border in Ireland. To avoid "hardening" the border, regulatory checks take place along the external coast of Ireland. However, disagreements over certain food products, which should be checked or, in some cases, barred from entering a non-EU country (e.g., chilled meats) generated controversy between the governments. After an allotted grace period, the UK and EU were unable to agree on how to resolve the problem. Openly refusing to follow the protocol, the UK continued to ship sausages and other chilled meats across the border. As a result, the EU launched legal actions against the UK for violating the protocol. Thus, it appears that the complex, tumultuous, and situated political geographies on the island of Ireland will continue to be an integral part of Brexit discussions.

Lessons learned from events explored in this book give us insight into how the Irish border conflict may play out across various pathways of change and makes sense of the geopolitical fluidity of the moment. By understanding the complicated and often tumultuous history, we understand there are no optimal solutions, as any path forward will upset some group or organization. No issue is as pressing and loaded with implications as the aftermath of Brexit.

In July 2021, British PM Boris Johnson and his chief negotiator with the EU, David Frost, published a document expressing their concerns over the Northern Ireland Protocol. If the UK forces the EU to relinquish the border requirement, it threatens the political trust established between the two governments as the document argues for the need to renegotiate the protocol.

Such action could result in potential trade sanctions against the UK for rejecting global trade norms. Furthermore, reopening border discussions threatens geopolitical relationships (e.g., the EU as the international guarantor of the 1998 peace agreement), could result in the return of a hard border in Ireland (which would be disastrous for the Republic), and erode mutual trust between the UK and the EU. Some have argued it threatens the future of the British union, as disagreements over the protocol are encouraging many within Northern Ireland to demand a referendum on the future of the region's union with the UK. When questioned about this possible development, Ireland's deputy PR, Leo Varadkar suggested that the "reunification of Ireland could occur in his lifetime" (Leahy

2021). Clearly, the role of Ireland has become a central factor in EU and UK relations.

If the UK caves to EU and court of justice pressure, thus becoming a "rule taker instead of a rule maker" it would represent a blow to Brexiters and the Johnson administration. Maintaining the protocol would also dramatically disrupt commerce between Northern Ireland and the rest of the UK, upsetting unionists/loyalists (thereby threatening the 1998 peace agreement), as well as impacting the region's business investment practices.[3]

Early evidence suggests the emergence of greater economic unity in Ireland than existed before the UK signed the protocol. This could continue to encourage the increase of exports from the Republic, thereby stimulating "all-Ireland" economic trade. If this trend were to last, it could bolster the prospect of the political reunification of Ireland. Economic integration requires the Republic to consider its options, given that it is not only part of the EU's Single Market but also fundamentally interested in protecting the peace in Northern Ireland. And history tells us that any moves toward integration would certainly lead to actions by loyalists, who are always looking for signs of the dissolution of Northern Ireland's union.

What is clear is that the uncertainty surrounding the border in Ireland is generating fear. As the Twitter sensation @BorderIrish, speaking as an anthropomorphized line, explained:

> the instability and fear, the sense of betrayal at the unravelling of a previously given guaranteed, which Brexit brings to this place. You know, a few years ago, before the Brexit thing, I imagined that whoever is in charge of these things would say to me one day, "Border, you've done a grand job, but we've no longer any need for you. We're letting you go." Instead, I wonder whatever happened to statecraft, that mixture of idealism and pragmatism that asserts a viewpoint and implicitly accepts the existence of another. When I look out, I see people of all kinds: good, caring, thoughtful, loving people; people who are open to change; and people who aren't . . . After nearly 100 years of being a border, the most important thing I've learnt is this: borders are the most cowardly form of human interaction. Opening yourself up to strangers, opening yourself up to the new and the unknown and the unexpected – that's bravery.
>
> *(2019, 247 & 249)*

Despite these serious concerns regarding the future of the Irish border, there is space for hope and continued efforts for reconciliation across various scales. While the island of Ireland continues to face challenges, some of which have not yet been

3 The COVID-19 pandemic also complicates the ability to separate new and emerging health protocols from the impact of Brexit (e.g., Dodds et al. 2020).

defined, many of its diverse inhabitants are increasingly utilizing space for forging peace, reconciliation, and repositioning former adversarial relationships. Various cross-border and cross-community peace initiatives continue to make inroads into previously divided communities and perceptions. Indeed, as this book has demonstrated, space is an integral part of political and cultural relations, perhaps best described by Holocaust survivor Victor Frankl when he stated:

> Between stimulus and response there is a space.
> In that space is our power to choose our response.

There is hope that the inhabitants of the island of Ireland will continue to employ space within their own situated geographies for future positive transformative endeavors both at home and far beyond national borders.

References

Anderson, Nicola. 2016. "How the Queen Brought History to Life during Momentous Irish Visit." *Irish News*, 2016.

@BorderIrish. 2019. *I Am the Border, So I Am*. New York, NY: HarperCollins.

Bowman, Timothy. 2014. "Ireland and the First World War." In *Oxford Handbook of Modern Irish History, 1600–2000*. Oxford: Oxford University Press.

Bunbury, Turtle. 2020. *Ireland's Forgotten Past: A History of the Overlooked and Disremembered*. London: Thames & Hudson.

Byrne, Elaine. 2014. "The Forgotten Irish Soldiers Who Fought for Britain in the First World War." *The Guardian*, 2014. www.theguardian.com/world/2014/apr/05/irish-soldiers-who-fought-for-britain.

Coakley, John. 2021. "Is a Middle Force Emerging in Northern Ireland?" *Irish Political Studies* 36 (1): 29–51.

DFA. 2021. "Ireland Takes up Seat on UN Security Council for 2021–2022 Term." *An Roinn Gnóthaí Eachtracha (Department of Foreign Affairs)*, January 1, 2021. www.dfa.ie/news-and-media/press-releases/press-release-archive/2021/january/ireland-takes-up-seat-on-un-security-council-for-2021–2022-term.php.

Dodds, Klaus, Vanesa Castan Broto, Klaus Detterbeck, Martin Jones, Virginie Mamadouh, Maano Ramutsindela, Monica Varsanyi, David Wachsmuth, and Chih Yuan Woon. "The COVID-19 Pandemic: Territorial, Political and Governance Dimensions of the Crisis." *Territory, Politics, Governance* 8 (3): 289–98.

Europa.eu/European-union. 2021. https://european-union.europa.eu/index_en

Higgins, Michael. 2011. "Presidential Acceptance Speech." Speech.

Johnson, Nuala C. 2003. *Ireland, the Great War and the Geography of Remembrance*. Vol. 35. Cambridge: Cambridge University Press.

Kelly, Fiach. 2019. "Varadkar Tells Johnson New Brexit Deal 'Not Going to Happen.'" *Irish Times*, July 24, 2019.

Leahy, Pat. 2021. "Leo Varadkar Believes Irish Unification 'Can Happen in My Lifetime.'" *Irish Times*, June 15, 2021. www.irishtimes.com/news/politics/leo-varadkar-believes-irish-unification-can-happen-in-my-lifetime-1.4594348.

McCormack, Jayne. 2019. "Boris Johnson. What Will the New Prime Minister Mean for NI?" *BBC*, July 23, 2019. www.bbc.com/news/uk-northern-ireland-49082366.

McGinnity, Frances, Eamonn Fahey, Emma Quinn, Samantha Arnold, Bertrand Maitre, and Philip O'Connell. 2018. *Monitoring Report on Integration 2018. ESRI Report*. Dublin: The Economic and Social Research Institute and Department of Justice and Equality.

Nolan, Paul. 2017. "Two Tribes: A Divided Northern Ireland." *The Irish Times*, April 1, 2017. www.irishtimes.com/news/ireland/irish-news/two-tribes-a-divided-northern-ireland-1.3030921.

INDEX

Note: Page numbers in *italics* indicate a figure or map and page numbers in **bold** indicate a table on the corresponding page. Page numbers followed by "n" indicate a note.

Milton Keynes UK
Ingram Content Group UK Ltd.
UKHW020627121223
434203UK00012B/139